THE LIBRARY
ST. MARY'S COLLEGE OF MARYLAND
ST. MARY'S CITY, MARYLAND 20686

Comparative Physiology of Osmoregulation in Animals

Frontispiece. Thomson's Gazelle. Small antelopes like the gazelles and dik-dik can live in dry areas without water and, in addition, their kidneys can excrete a urine with a maximal osmolality ranging between 3000 and 4500 mOsm/kg H_2O.

Comparative Physiology of Osmoregulation in Animals

Edited by

G. M. O. MALOIY

*Department of Animal Physiology
University of Nairobi,
P.O. Box 30197,
Nairobi, Kenya*

Volume 2

1979

ACADEMIC PRESS
London   New York   San Francisco
A Subsidiary of Harcourt Brace Jovanovich, Publishers

ACADEMIC PRESS INC. (LONDON) LTD
24–28 Oval Road
London NW1

U.S. Edition published by
ACADEMIC PRESS INC.
111 Fifth Avenue
New York, New York 10003

Copyright © 1979 by
ACADEMIC PRESS INC. (LONDON) LTD.

All Rights Reserved
No part of this book may be reproduced in any form by photostat, microfilm, or any other means, without written permission from the publishers

British Library Cataloguing in Publication Data
Comparative physiology of osmoregulation in animals
Vol. 2
1. Osmoregulation
I. Maloiy, GMO
591.1′9′212   QP90.6   77-93492
ISBN 0-12-467002-4

PRINTED IN GREAT BRITAIN BY
T. & A. CONSTABLE LTD., EDINBURGH

LIST OF CONTRIBUTORS

P. J. BENTLEY, *Department of Pharmacology, Mount Sinai School of Medicine, The City University of New York, Fifth Avenue and 100th Street, New York, NY 10029, USA.*

H. J. FYHN, *Institute of Physiology, University of Bergen, Bergen, Norway.*

W. V. MACFARLANE, *Department of Animal Physiology, Waite Agricultural Research Institute, University of Adelaide, Glen Osmond, South Australia.*

G. M. O. MALOIY, *Department of Animal Physiology, University of Nairobi, Nairobi, Kenya.*

R. L. MALVIN, *Department of Physiology, Medical School, University of Michigan, Ann Arbor, Michigan 48109, USA.*

M. PEAKER, *Physiology Department, The Hannah Research Institute, Ayr, Scotland.*

A. SHKOLNIK, *Department of Zoology, Tel-Aviv University, Tel-Aviv, Israel.*

L. P. SULLIVAN, *Department of Physiology, University of Kansas, Medical Centre, Kansas City, Kansas, USA.*

E. J. WILLOUGHBY, *Division of Natural Sciences and Mathematics, St Mary's College of Maryland, St Mary's City, Maryland 20686, USA.*

PREFACE

Living things have an impressive ability to adapt to extreme osmotic environments. In most of the invertebrates and some of the lower vertebrates considered in the previous volume, this is often an ability to withstand a wide range of changes in the internal environment. Rather than spend energy modifying their body fluids, they simply tolerate them. No such easy option is available for the higher vertebrates discussed in this volume, for their more highly evolved tissues perish unless maintained under very narrowly defined limits. Thus, for the subjects of this volume, adaptation to a hostile environment is generally a function of homeostatic systems which may be called upon to regulate body fluid concentrations under extreme conditions.

Heat and aridity, water deprivation, high insolation and convection with consequent high evaporation characterize hot deserts, but equally severe problems of water loss are faced by inhabitants of high altitudes. Body size becomes an important factor in reducing evaporative water loss but may itself be limited by scarcity of food. Food may also cause difficulties with ionic balance—for example, the potassium-rich, sodium-poor herbivore diet, or the acidosis-inducing carnivore diet. Adaptations for handling food may further stress the homeostatic regulation of body fluids; the ruminant stomach, although a superb example of makeshift engineering, appears to be in precarious osmotic balance with the rest of the body such that a minor rise in rumen osmolality will cause a catastrophic shift of water out of the interstitial fluid. The ability of animals to survive such perils demonstrates how reliable osmotic and ionic regulation has become since the first cell surrounded itself with a semi-permeable membrane.

There are, of course, limits to the conditions that can be tolerated. Distribution of animals is determined partly by the ability of the ion transport mechanisms of the gut nasal salt gland in birds and kidney to maintain constant internal conditions. In this area, where homeostatic mechanisms reach their limit, ecology and physiology become complementary. It is hoped that this volume may be of use to workers from both disciplines.

This volume includes contributions on osmoregulatory problems facing terrestrial animals such as birds, rodents, carnivores, herbivores, primates and Australian marsupials and monotremes.

During the preparation of this volume, I have benefited from the advice and suggestions from a number of colleagues. I am particularly thankful to Drs O. S. Bamford, V. A. Langman, E. T. Clemens and

J. M. Z. Kamau. They are not, however, responsible for any errors which might become apparent in this volume.

My own research on Ionic and Osmotic regulation has been supported financially by the Leverhulme Trust Fund, the Wellcome Trust and the National Geographic Society.

I would like to take this opportunity to acknowledge the excellent cooperation I have received from the staff of Academic Press. Mrs Fay Frost and Mrs Mary Kamano have provided secretarial services with charm, skill and patience.

Nairobi   G. M. O. MALOIY
June, 1979

CONTENTS

List of Contributors v

Contents vii

Contents of Volume 1 xi

Chapter 1. Birds—E. J. Willoughby and M. Peaker 1

Chapter 2. Australian Marsupials and Monotremes—P. J. Bentley 57

Chapter 3. Rodents—H. J. Fyhn 95

Chapter 4. Carnivores—R. L. Malvin 145

Chapter 5. Mammalian Herbivores—G. M. O. Maloiy, W. V. Macfarlane and A. Shkolnik 185

Chapter 6. Primates—L. P. Sullivan 211

Subject Index 233

CONTENTS OF VOLUME 1

Chapter 1. Principles of Ion Transport in Kidney Tubules—G. GIEBISCH 1

Chapter 2A. Marine and Brackish Water Animals—D. H. SPAARGAREN 83

Chapter 2B. Fresh Water Invertebrates—P. GREENAWAY 117

Chapter 3. Terrestrial Invertebrates Other Than Insects—L. H. MANTEL 175

Chapter 4. Insects—B. J. WALL and J. L. OSCHMAN 219

Chapter 5. Amphibians—R. H. ALVARADO 261

Chapter 6. Fish—D. H. EVANS 305

Chapter 7. Reptiles—J. E. MINNICH 391

SUBJECT INDEX 643

1. Birds

E. J. WILLOUGHBY and M. PEAKER*

St Mary's College of Maryland, USA and
*Institute of Animal Physiology, Babraham, Cambridge, England

I.	Introduction	1
II.	Pathways and Control of Water Balance	2
	A. Water Turnover	3
	B. Evaporative Water Loss	3
	C. Water Loss Through the Kidneys	9
	D. Water Loss Through the Nasal Salt Gland	17
	E. Water Budgets	19
III.	Pathways and Control of Ionic Balance	23
	A. Renal Structure and Function	23
	B. Post-renal Modification of Urine	34
	C. Nasal Salt Glands	36
IV.	Summary	48
	References	49

I. INTRODUCTION

The birds, class Aves, are a terrestrial group with representatives that successfully inhabit all oceanic and continental areas of the globe. They do so largely through their highly evolved capacities for osmotic and ionic regulation.

In the past 30 years in particular, physiologists have been studying the comparative physiology of avian osmoregulation, especially regarding adaptations to osmotically stressful desertic and marine environments, and much of this work has been reviewed lately (Chew, 1961; Bartholomew and Cade, 1963; Cade, 1964; Holmes, 1965; Sturkie, 1965; Dawson and Bartholomew, 1968; Dantzler, 1970;

* Present address: *Department of Physiology, The Hannah Research Institute, Ayr, Scotland.*

Shoemaker, 1972). Here we shall review the principles of avian osmotic and ionic regulation, and discuss information on physiological processes and adaptive patterns that is emerging from continuing studies. For convenience, we have organized the subject into two main sections, one on water metabolism, the other on excretory processes.

II. PATHWAYS AND CONTROL OF WATER BALANCE

Osmoregulation depends on physiological processes that promote a balance between loss and gain of body water. Water is gained through the gut as preformed water in food and drink and by intracellular oxidation of hydrogen in foodstuffs. It is lost by evaporation from body surfaces, through the kidney as urine, through the gut in the faeces, and sometimes through nasal salt secretion.

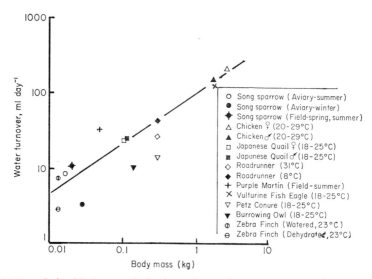

Fig. 1. The relationship between body size and rate of water turnover as determined with injections of HTO or $D_2^{18}O$. The regression is the equation of Ohmart et al. (1970a), $R = 105 W^{0.69}$. Sources of data: song sparrow (*Melospiza melodia*), Stephenson, 1974; burrowing owl (*Speotyto cunicularia*), Japanese quail (*Coturnix coturnix*), Petz conure (*Aratinga canicularis*) and vulturine fish eagle (*Gypohierax angolensis*), Chapman and McFarland (1971); chicken (*Gallus gallus*), Chapman and Black (1967); purple martin (*Progne subis*), calculated from data of Utter and LeFebvre (1973); zebra finch (*Poephila guttata*), Skadhauge and Bradshaw (1974). Redrawn from Stephenson (1974) with addition of zebra finch. Dehydrated zebra finches were deprived of drinking water during the test; watered zebra finches drank water *ad libitum*. The reduced rate of water turnover of captive song sparrows in winter compared with other groups is related to reduced environmental temperatures (Stephenson, 1974). However, the opposite relationship appears to apply to the roadrunner.

A. WATER TURNOVER

Recently, water turnover rates for several species have been determined by measuring the rate of disappearance of tritiated water or $D_2{}^{18}O$ following injection into captive and free-living birds. The object of such studies usually has been to measure the overall rate of water metabolism in order to reveal adaptive patterns. Thus, birds adapted to arid habitats might be expected to have lower rates of water turnover than those occupying mesic habitats, etc. The data are summarized in Fig. 1. Ohmart et al. (1970a) calculated the relationship between body mass and water turnover for their limited sample as approximating the equation $R = 105W^{0.69}$, where R is rate in ml day^{-1}, and W is the mass in kg. These rates of water metabolism are in general agreement with those measured by methods other than with isotopic tracers (Table VIII). However, the lack of standardization of the environmental and physiological conditions during measurements makes generalizations risky, and comparisons between different species should be made cautiously.

B. EVAPORATIVE WATER LOSS

In birds, a major part of the water loss is by evaporation. Their high metabolic rate, connected with homeothermy, small body mass, and flight, promotes high rates of evaporation through the respiratory surfaces; and their small body mass also gives relatively large surface areas for evaporation from the skin. Physiological adaptations that reduce evaporation would promote survival in habitats where preformed water is scarce or inaccessible.

1. *Rates of Evaporative Water Loss*

Although birds regularly vary their rates of evaporation for temperature regulation, one sees interspecific and intraspecific differences in evaporation rates that reflect differences in body mass and shape, or adaptation to dehydrating conditions, or both. Crawford and Lasiewski (1968) derived the relationship $\log E = \log 0.432 + 0.585 \log W$, where E is evaporative water loss in g H_2O day^{-1}, and W is body mass in g, based on measurements from resting birds weighing from 3 g (hummingbirds) to 100 kg (ostrich) at temperatures at or below thermal neutrality. As Crawford and Lasiewski pointed out, the values on which this equation is based are from various sources, measured by various methods under varying conditions of ambient water vapour pressure, time of day, and nutritional and metabolic state of the experimental subjects, and therefore should be considered preliminary. Table I presents some representative values of evaporative water loss, and Fig. 2 depicts these values plotted against body mass on logarithmic coordinates to show the general relationship between body mass and rate of evaporation.

TABLE I

Rates of evaporation of birds resting at 23 to 25°C

Species	Body mass (g)	Evaporation rate (% body wt day^{-1})	Reference
Caliope hummingbird *Stellula caliope*	3·0	33·6	Lasiewski (1963, 1964)
Anna's hummingbird *Calypte anna*	4·8	18·3	Lasiewski (1963, 1964)
White-rumped waxbill *Estrilda troglodytes*	6·0	28·8	Cade et al. (1965)
House wren *Troglodytes aedon*	9·7	36·0	Kendeigh (1939)
Silverbill *Lonchura malabarica*[a]	10·9	16·8	Willoughby (1969)
Zebra finch *Poephila guttata* (= *Taeniopygia castanotis*)[a]	11·5	20·6	Cade et al. (1965)
Red-breasted nuthatch *Sitta canadensis*	11·2	12·0	Mugaas and Templeton (1970)
Grey-backed finch-lark *Eremopteryx verticalis*[a]	15·1	9·6	Willoughby (1968)
Stark's lark *Spizocorys starki*[a]	15·6	12·0	Willoughby (1968)
Sage sparrow *Amphispiza belli nevadensis*	18	13·3	Moldenhauer and Wiens (1970)
House finch *Carpodacus mexicanus*	18·8	17·2	Bartholomew and Dawson (1953)
White-crowned sparrow *Zonotrichia leucophrys*	23·3	13·6	Bartholomew and Dawson (1953)
House sparrow *Passer domesticus*	25	14·4	Kendeigh (1944)
Horned lark *Eremophila alpestris*	26	15·0	Trost (1968)
Budgerigar *Melopsittacus undulatus*[a]	35	5	Greenwald et al. (1967)
Bourke parrot *Neophema bourki*	35·3	7·7	Dawson (1965)

TABLE I (contd.)

Species	Body mass (g)	Evaporation rate (% body wt day $^{-1}$)	Reference
Rufous-sided towhee *Pipilo erythrophthalmus*	35·4	13·3	Bartholomew and Dawson (1953)
Brown-headed cowbird *Molothrus ater*	36·2	13	Lustick (1970)
Abert's towhee *Pipilo aberti*	38·2	7·7	Bartholomew and Dawson (1953)
Brown towhee, *Pipilo fuscus*	39·3	6·0	Bartholomew and Dawson (1953)
Mockingbird *Mimus polyglottos*	39·6	5·4	Bartholomew and Dawson (1953)
Poorwill *Phalaenoptilus nuttalii*	40	6·0	Bartholomew et al. (1962)
Lesser nighthawk *Chordeilis acutipennis*	40·2	7·5	Bartholomew and Dawson (1953)
Rock parrot *Neophema petrophila*	40·7	8·1	Dawson (1965)
Loggerhead shrike *Lanius ludovicianus*	40·8	5·8	Bartholomew and Dawson (1953)
Inca dove *Scardafella inca*	42	7·5	MacMillen and Trost (1967)
Speckled mousebird *Colius striatus*	52·5	5·0	Bartholomew and Trost (1970)
California thrasher *Toxostoma redivivum*	74·7	4·7	Bartholomew and Dawson (1953)
Cockatiel *Nymphicus hollandicus*	80·7	4·2	Dawson (1965)
Screech owl *Otus asio*	101·3	3·9	Bartholomew and Dawson (1953)
Mourning dove *Zenaida macroura*	118·7	2·5	Bartholomew and Dawson (1953)
Twenty-eight parrot *Platycercus zonarius*	137·0	2·6	Dawson (1965)
Burrowing owl *Speotyto cunicularia*	138·0	4·8	Coulombe (1970)
Gambel's quail *Lophortyx gambelii*	148·8	3·2	McNabb (1969)

TABLE I (contd.)

Species	Body mass (g)	Evaporation rate (% body wt day^{-1})	Reference
California quail *L. californicus*	149·4	3·5	McNabb (1969)
Bobwhite *Colinus virginianus*	180·5	3·1	McNabb (1969)
Roadrunner *Geococcyx californianus*	284·7	2·4[b]	Calder and Schmidt-Nielsen (1967)
Rock dove *Columba livia*	314·6	2·4[b]	Calder and Schmidt-Nielsen (1967)
White-tailed ptarmigan *Lagopus leucurus*	326	4·8	R. E. Johnson (1968)
Chicken *Gallus gallus*	3200	2·2	Dicker and Haslam (1972)
Goose *Anser anser*	4800	1·5	Benedict and Lee (1937)
Ostrich *Struthio camelus*	84 000	0·6	Schmidt-Nielsen et al. (1969)

[a] Seed-eating species from desertic habitats that can withstand water deprivation better than most birds.
[b] Values estimated from data in graphs.

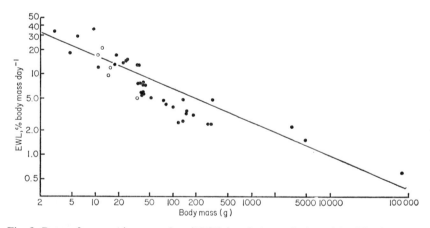

Fig. 2. Rates of evaporative water loss (EWL) in relation to body weight. The data come from Table I. ●, species not known to be especially resistant to dehydration; ○, seed-eating species from desertic environments that can withstand water deprivation better than most birds. The regression line is the equation of Crawford and Lasiewski (1968) converted to give E as $\%W$ day^{-1} or $\log E = \log 43·2 - 0·415 \log W$, where W is body weight in g.

2. Comparison of Cutaneous and Respiratory Evaporation

Water evaporates from all body surfaces in contact with air. For many years investigators assumed that most evaporation occurs from the moist surfaces of the respiratory tract, and that the comparatively dry and glandless skin contributes insignificantly to the total. This may be true for some birds that are actively promoting evaporative water loss to cool themselves in hot environments, but it is not always so in

TABLE II

Cutaneous evaporation as a proportion of total evaporation in birds resting at 30 to 35°C

Species	Body mass (g)	Cutaneous evaporation (% of total)	Reference
Zebra finch			
Poephila guttata	12·5	62·9	Bernstein (1971)
Budgerigar			
Melopsittacus undulatus	31·6	58·9	Bernstein (1971)
Painted quail			
Excalifactoria chinensis	42·3	44·7	Bernstein (1971)
Village weaver			
Ploceus cucallatus	42·6	50·8	Bernstein (1971)
Poorwill			
Phalaenoptilus nuttalii	43·2	51·3	Lasiewski et al. (1971)
Roadrunner			
Geococcyx californianus	274·2	51·0	Lasiewski et al. (1971)
Rock dove			
Columba livia	299·9	74	Smith (1969), cited in Lasiewski et al. (1971)

the absence of thermal stress. In six of seven species ranging from 12·5 to 300 g for which comparable data are available, the skin contributes more than half the total evaporation (Table II).

3. Control of Evaporative Water Loss

Several granivorous species native of dry habitats can reduce their total evaporation when drinking water is withheld or restricted, as an adaptive response to dehydrating conditions (Table III). Lee and Schmidt-Nielsen (1971) have shown that in the zebra finch the reduction in evaporation involves the skin rather than the respiratory tract. The mechanism of the reduction of cutaneous evaporation

TABLE III

Reduction of evaporative water loss (EWL) by small birds adapting to dehydrating conditions

Species	Body mass (g)	T$_A$ (°C)	Mean EWL hydrated[a] (mg g^{-1} h^{-1})	Mean EWL hydropenic[b] (mg g^{-1} h^{-1})	Reference
Zebra finch *Poephila guttata*	12	40	19·8	13·9	Calder (1964)
	11·5	35	9·5	5	Cade *et al.* (1965)
	11·5	40	21·5	8	Cade *et al.* (1965)
	12·0	25	6·9	4·1	Lee and Schmidt-Nielsen (1971)
Silverbill *Lonchura malabarica*	9·7	37	10·5	5	Willoughby (1969)
Stark's lark *Spizocorys starki*	15·6	24–30	5	3·5	Willoughby (1968)
Budgerigar *Melopsittacus undulatus*	35	40	16·5	12	Greenwald *et al.* (1967)

[a] Hydrated means birds were given drinking water *ad libitum*.
[b] Hydropenic means birds were kept without drinking water, or were restricted to 4% of their *ad libitum* drinking rate.

remains obscure. Nevertheless, any reduction in evaporation would benefit osmoregulation in a dry habitat.

C. WATER LOSS THROUGH THE KIDNEYS

The formation of urine to eliminate metabolic wastes is an important pathway of water loss. Birds minimize this loss by producing hyperosmotic urine, and by excreting uric acid as the principal end product of nitrogen metabolism. Since uric acid and its salts are only slightly soluble, they contribute little to the osmotic concentration of the urine after they have been precipitated in the urinary tract. This frees water for resorption, so most of the osmotic concentration of the urine results from other solutes, including various salts.

1. *Nitrogen Excretion*

Birds excrete nitrogen mostly as uric acid, ammonia and urea. The term uric acid here refers both to uric acid and its salts (urates), unless otherwise noted. The proportions of uric acid to urates can be expected to vary as the pH and ionic composition of the urine vary. Uric acid accounts for about 60 to 80% of urinary nitrogen in chickens (Sturkie, 1965). The majority of this uric acid is secreted into the renal tubule but the site of secretion is not known. It is likely that the principal site of secretion is the proximal convoluted tubule, as it is in snakes (Dantzler, 1973; Randle and Dantzler, 1973).

The proportion of uric acid appearing as a precipitate compared with that in a dissolved or colloidally suspended form in the liquid fraction of the urine varies with the total concentration of uric acid. In a copious, dilute urine, as little as 15 to 20% appears as an insoluble precipitate, while in a very concentrated urine as much as 93 to 97% is in the solid form (McNabb, 1974). Precipitation of uric acid occurs in the collecting tubules, and mucus secreted in them and along the ureteral branches, by acting as a lubricant, probably helps prevent clogging of the excretory passages (F. M. A. McNabb *et al.*, 1973; McNabb and Poulson, 1970). The precipitate consists of spheres 2 to 8 μm in diameter (Folk, 1969) which appear to consist mainly of uric acid with a smaller admixture of soluble materials (Lonsdale and Sutor, 1971). The processes that form these spheres are unknown, but the spheres are unstable and outside the body they gradually break down into the normal type of uric acid crystals.

Besides freeing osmotically unobligated water for resorption, solid uric acid contributes to water conservation in another way. A significant part of the Na^+, K^+, Ca^{2+} and Mg^{2+} in the whole urine can occur as undissolved urates or in a form otherwise bound to the precipitated uric acid. In a study comparing urate and cation excretion of roosters fed high or low protein diets and tap water or salt water to drink, R. A. McNabb et al. (1973) observed that from 3 to 75% of urinary Na^+ and from 8 to 34% of urinary K^+ were associated with the precipitated urates. The largest percentages occurred in roosters eating a high protein diet and drinking tap water. In another series of experiments on roosters drinking tap water or salt water and eating low protein diet, McNabb and McNabb (1977) found 24 to 32% of urinary Ca^{2+} and 17 to 24% of urinary Mg^{2+} associated with the precipitated urates. Again, the higher percentages were observed in birds drinking tap water. These precipitated cations thus did not contribute to the osmotic concentration of the urine, and therefore would have freed additional water for resorption. These investigators see reason to believe that the cations are not all present in the form of urate salts, but instead probably occur in the urinary spheres principally as layers of cations electrostatically sandwiched between layers of uric acid dihydrate. The report by Minnich (1972) that NH_4^+ is bound with precipitated urates in several species of lizards and snakes and a tortoise would suggest that this cation may also be bound with urates in avian urine; and McNabb (1974) did observe that about 15% of the urinary NH_4^+ was present in urinary precipitates in roosters, although he did not consider this to be a significant fraction of the excreted NH_4^+.

2. Rate of Urine Production

Birds vary their rate of urine production over a wide range according to the osmotic requirements associated with various levels of water intake. Table IV lists rates of ureteral urine discharge associated with various treatments in the chicken, duck and budgerigar. In each species, urine output from the ureters can vary by a factor of three or more. Where data on intake are available for the chicken and budgerigar, it can be seen that urine output is between $\frac{1}{2}$ and 1 times intake. The situation for the duck appears more complicated, and this may be owing to water loss through nasal salt secretion (the chicken and budgerigar lack nasal salt glands). The control of the rate of urine production will be discussed later in Section III.

3. Osmotic Concentration of Urine

The varying rates of secretion of urine are associated with varying osmotic concentrations. Unlike reptiles, but like mammals, birds in a

TABLE IV

Rates of urine production under varying degrees of water and salt loading[a]

Species	Body mass (kg)	Treatment	Fluid intake (ml kg^{-1} h^{-1})	Urine production (ml kg^{-1} h^{-1})	Reference
Chicken, *Gallus gallus*	2·1	Food and water *ad libitum*	—	1·76	Hester *et al.* (1940)
	1–3	Single oral dose of H$_2$O, 9% of body wt	—	17·9	Skadhauge and Schmidt-Nielsen (1967)
		Deprived of water for 36 h	—	1·08	Skadhauge and Schmidt-Nielsen (1967)
	1·44	Intravenous infusion of hypoosmotic solution	15–23[b]	12·5	Ames *et al.* (1971)
	3·2	Food and water *ad libitum*	1·26	0·53	Dicker and Haslam (1972)
Duck, *Anas platyrhynchos*[c]	2·1	Intermittent doses of distilled water by stomach	7·9	7·8	Holmes (1965)
		Intermittent doses of hypertonic saline[d]	7·9	0·57	Holmes (1965)
Duck, *Anas platyrhynchos*[c]	2·2	Fresh water *ad libitum*	—	2·4	Holmes *et al.* (1968)
		Fresh water *ad libitum* plus intravenous infusion of isoosmotic saline	0·84–1·68[b]	3·4	Holmes *et al.* (1968)
Budgerigar, *Melopsittacus undulatus*	0·029–0·046	Deprived of drinking water, intravenous infusion of isotonic saline	1·97[b]	1·98	Krag and Skadhauge (1972)
		Intravenous infusion of hypotonic saline (40 mosmol)	13·1[b]	6·3	Krag and Skadhauge (1972)

[a] Unless otherwise noted, urine was collected from the ureteral openings by techniques other than cannulation.
[b] Rate of intravenous infusion.
[c] Urine was collected as birds released it intermittently from the cloaca.
[d] Saline doses given by stomach tube through esophagus.

TABLE V

Maximum urine to plasma osmotic ratios in birds deprived of water or given strongly hyperosmotic saline solutions

Species	Treatment	Mean concentration (mosmol) Plasma	Mean concentration (mosmol) Urine	U/P	Reference
		Birds lacking nasal salt glands			
Chicken, *Gallus gallus*	Water deprivation	341	538	1·58	Skadhauge and Schmidt-Nielsen (1967)
Turkey, *Meleagris gallopavo*	Hyperosmotic saline	—	553	1·6[a]	Skadhauge and Schmidt-Nielsen (1967)
	Water deprivation	—	492	1·4[a]	Skadhauge and Schmidt-Nielsen (1967)
Bobwhite, *Colinus virginianus*	Hyperosmotic saline	347	605	1·7	McNabb (1969b)
	Water deprivation	395	643	1·6	McNabb (1969b)
California quail, *Lophortyx californicus*	Hyperosmotic saline	374	642	1·7	McNabb (1969b)
	Water deprivation	338	669	2·0	McNabb (1969b)
	Hyperosmotic saline	—	—	2·0	Carey and Morton (1971)
Gambel's quail, *L. gambelii*[b]	Hyperosmotic saline	348	884	2·5	McNabb (1969b)
	Water deprivation	337	669	2·0	McNabb (1969b)
	Hyperosmotic saline	—	—	2·4	Carey and Morton (1971)
Mourning dove, *Zenaida macroura*	Hyperosmotic saline	408·5	544·0	1·33	Smyth and Bartholomew (1966)
Senegal dove, *Streptopelia senegalensis*	Water deprivation	379	661	1·74	Skadhauge (1974b)
Crested pigeon, *Ocyphaps lophotes*[b]	Water deprivation	370	655	1·77	Skadhauge (1974b)
Budgerigar, *Melopsittacus undulatus*[b]	Water deprivation	365	848	2·33	Krag and Skadhauge (1972)
Galah, *Cacatua roseicapilla*[b]	Water deprivation	388	973	2·5	Skadhauge (1974a)
	Water deprivation	400	982	2·5	Skadhauge (1974b)

Species	Treatment				Reference
Singing honey eater, *Meliphaga virescens*	Water deprivation		384	2.41	Skadhauge (1974b)
Red wattle bird, *Anthochaera carunculata*	Water deprivation		388	2.36	Skadhauge (1974b)
Zebra finch, *Poephila guttata*[b]	Water deprivation		361	2.78	Skadhauge (1974b)
Sage sparrow, *Amphispiza belli*[b]	Hyperosmotic saline		314.5	2.04	Moldenhauer and Wiens (1970)
Savannah sparrow, salt-marsh race, *Passerculus sandwichensis beldingi*	Hyperosmotic saline		450[c]	4.5	Poulson and Bartholomew (1962a)
Savannah sparrow, fresh marsh race, *P. s. brooksi*	Hyperosmotic saline		440[c]	2.2	Pouslon and Bartholomew (1962a)
House finch, *Carpodacus mexicanus*	Hyperosmotic saline		360[c]	2.36	Poulson and Bartholomew (1962b)
Emu, *Dromaius novaehollandiae*[b]	Water deprivation		337	1.36	Skadhauge (1974b)
Kookaburra, *Dacelo gigas*	Water deprivation		348	2.71	Skadhauge (1974b)
Birds having nasal salt glands					
Duck, *Anas platyrhynchos*	Hyperosmotic saline	—	462	1.3[a]	Skadhauge and Schmidt-Nielsen (1967)
	Hyperosmotic saline	313	444	1.42	Helmes *et al.* (1968)
Pelican, *Pelecanus erythrorhynchos*	Water deprivation	—	580	1.7[a]	Calder and Bentley (1967)
Roadrunner, *Geococcyx californianus*[b]	Water deprivation	—	593	1.7[a]	Calder and Bentley (1967)
Ostrich, *Struthio camelus*[b]	Water deprivation	290[c]	760[c]	2.6	Louw *et al.* (1969)

[a] Value calculated on the assumption that plasma was 350 mosmol.
[b] Species native to desertic habitats.
[c] Values estimated from graph.

dehydrated state can produce urine that is hyperosmotic to the plasma. Generally, the maximum osmotic urine to plasma ratio (U/P) attained approaches 2. Table V presents values for maximum U/P ratios in several species from both dry and mesic habitats. Of 9 species native to desertic regions, 6 have a maximum U/P greater than 2, while of 8 species from mesic habitats, 4 have a maximum U/P greater than 2. The salt marsh savannah sparrow, native to coastal salicornia marshes and alkaline sloughs in southern California, produces urine with the highest known U/P. This is undoubtedly an adaptation to the osmotically difficult habitat, in which rainfall does not occur during the hot summer months, and where the main source of free water is the sea water or the salty sap in the halophytic marsh vegetation. Even the subspecies that breeds in fresh water marshes and meadows occupies the salt marsh habitat of the former in winter, although it is not then restricted to this habitat (Grinnell and Miller, 1944), and its U/P of 2·2 may be adaptive to marshes with brackish water in its winter range.

Skadhauge (1974b) pointed out that the red wattle bird and singing honeyeater (family Melliphagidae) are both active in daytime during the hot summer months near Perth, Western Australia, where their U/P ratios were measured. Such behaviour would subject these birds to thermal conditions requiring evaporative cooling, so the good concentrating ability of their kidneys may be an adaptation that helps make water available for thermoregulation.

In comparing xerophilic with mesophilic species in Table V, one should consider it likely that the sample is biased toward birds having high U/P, as that is what many investigators sought. Therefore, this limited sample can only be taken as being suggestive that birds from arid habitats do have enhanced urine concentrating abilities compared with birds of mesic habitats.

4. *Resorption of Water in Cloaca and Rectum*

Because urine passes into the cloaca and is normally held there for some time before it is discharged, and because birds can not produce a urine as concentrated as that of mammals, it has long been argued that the cloaca plays an important part in avian water balance by resorbing urinary water. Many experiments have been done on chickens to test this hypothesis. They generally involved surgical procedures that separated the ureteral orifices from the cloaca in experimental birds, and left the ureters and cloaca intact in the controls, but with the rectum disconnected or plugged so that the collected urine would not be contaminated with faeces (Sturkie, 1965; Skadhauge, 1973). From these studies one can conclude that the cloaca by itself plays no significant role in resorption in the chicken (Dixon, 1958; Nechay and Carmen Lutherer, 1968).

The recent demonstrations that ureteral urine rapidly moves from the cloaca into the large intestine, or rectum, in chickens and ducks (Akester et al., 1967; Skadhauge, 1968; Nechay et al., 1968) and the roadrunner (Ohmart et al., 1970b) have rekindled experimentation on post-renal modification of urine, now including the action of the intestinal mucosa in these processes. Currently, references to the cloaca as a modifier of urine generally include the action of the rectum, so, unless otherwise noted, we shall use the term cloaca in this broad sense.

Studies by Skadhauge (1967, 1974a) and Bindslev and Skadhauge (1971a, 1971b) have gone far to elucidate the osmotic and ionic processes in the avian cloaca and large intestine that tend to alter composition of the urine. They used *in vivo* perfusion of the coprodaeum and large intestine of anaesthetized birds (chicken, galah) with solutions of varying osmotic and ionic composition to measure rates and direction of water and ion flow across the cloacal epithelium. These studies have shown that there can be a net absorption of water in the chicken cloaca with perfusion fluids ranging from hypoosmotic to 60–80 mosmol higher than plasma (Skadhauge, 1967; Bindslev and Skadhauge, 1971b), apparently somewhat higher in dehydrated birds than in normally hydrated ones at cloacal perfusion rates of 5–9 ml h^{-1}. In the same experiments, when perfusion rates were reduced to 0·8–1·0 ml kg^{-1} h^{-1}, comparable to the rate of urine flow in dehydrated birds, the results led to an estimate that water absorption could occur from the incoming perfusion fluids that exceeded plasma osmalality by up to about 175 mosmol in dehydrated birds.

The absorption of water against the osmotic gradient can be attributed to a solute-linked water transport which probably results from the active transport of Na$^+$ by the cloacal epithelium. This water flow in normally hydrated chickens amounted to $1·07 \pm 0·52$ μl H$_2$O μEq^{-1} Na^{+-1} absorbed (mean ± s.d.), and in dehydrated birds, it averaged $1·48 \pm 0·52$ μl μEq^{-1}, but the difference is not statistically significant (Bindslev and Skadhauge, 1971b). In the dehydrated galah, a xerophilous, Australian seed-eating parrot, Skadhauge (1974a) estimated the solute-linked water transport to be $5·0 \pm 0·4$ μl μEq^{-1}. The adaptive significance of this difference may be that the dehydrated galah can store urine, which is more hyperosmotic than that of a dehydrated chicken (Table V), in the cloaca without a significant loss of water to it by osmosis, owing to the compensating uptake of moderately hyperosmotic absorbate (385 mEq l^{-1}) (Skadhauge, 1974a). This contrasts with the much higher absorbate concentration of the dehydrated chicken, estimated at about 1350 mEq l^{-1} (Bindslev and Skadhauge, 1971b).

An additional finding in these studies was that the permeability to osmotic transport of the cloacal epithelium was greater in the mucosal to serosal direction than in the reverse direction—that is, there was an osmotic rectification. The osmotic permeabilities as μl (kg h mosmol)$^{-1}$ in normally hydrated chickens were: serosal to mucosal = $3 \cdot 2 \pm 0 \cdot 54$ (mean ± s.d.), mucosal to serosal = $5 \cdot 8 \pm 0 \cdot 5$ and for dehydrated chickens they were: serosal to mucosal = $3 \cdot 6 \pm 0 \cdot 67$, mucosal to serosal = $10 \cdot 0 \pm 0 \cdot 75$ (Bindslev and Skadhauge, 1971a). These differences would promote osmotic uptake of water from hypoosmotic urine, but in comparison would retard osmotic dilution of hyperosmotic urine.

It is difficult to estimate the true significance of cloacal water resorption in the normal living bird from these perfusion studies. In an attempt to do so, Skadhauge and Kristensen (1972) devised a mathematical model of cloacal resorption based on a number of simplifying assumptions and incorporating such factors as rates of ureteral urine discharge, urine osmolality, osmotic permeability coefficients of the cloacal wall, rates of solute-linked water flow, and so on, for which measured values were available. The model was designed to predict volume changes of cloacal fluid resulting from an integration of the various measured parameters. It indicated that hyperosmotic urine may enter the cloaca without causing further water loss, and also that absorption of up to 14% of ureteral water can occur, but that the accompanying absorption of NaCl is so large that the total absorbate becomes hyperosmotic to plasma. These authors argued that such a process may benefit dehydrated desert birds, which can tolerate elevated plasma osmolality well, by helping to maintain plasma volume (Skadhauge and Kristensen, 1972). Another point to consider is that a resorption of salt and water from the urine of dehydrated birds should permit a continuing production of urine by the kidney sufficient to eliminate uric acid, which can then be concentrated in the cloaca with the resorption of water (Skadhauge, 1973).

A recent study by Dicker and Haslam (1972) on water balance in chickens with exteriorized ureters versus intact controls when both groups were feeding and drinking *ad libitum* gives the best picture available on the extent of water resorption in the cloaca over a period of days. The birds with exteriorized ureters excreted 40% more water through urine and faeces than did the controls, which they balanced by a corresponding increase in the amount of water drunk. Thus normal discharge of urine into the cloaca permitted these birds to conserve water lost by this pathway. The amount of water resorbed was equal to about one-fifth of the total water intake. Dicker and Haslam suggested that the absorption takes place largely by osmosis from hypoosmotic urine, which the chicken temporarily produces after feeding or drinking. Thus the daily urine production consists of two fractions, one of large flow rate and low osmolality (hypoosmotic) produced after eating or drinking, and one of low flow rate and high osmolality (hyperosmotic).

The role of the cloaca in ionic balance, especially in conjunction with the activity of the salt-secreting nasal gland, will be discussed in Section III. B.

D. WATER LOSS THROUGH THE NASAL SALT GLAND

The ionic concentration of the secretion of the nasal gland in marine and in some terrestrial birds is high (0·4 to 1·2 M NaCl, see below) and the glands, by excreting such concentrated solutions, serve as a mechanism to gain osmotically free water. The rate of secretion can also be high—up to about 2 ml (g tissue min)$^{-1}$—so that the glands constitute a significant avenue of water loss. However, it is unlikely that the glands secrete at a high rate continuously throughout the day, and Holmes et al. (1968) calculated that in ducks adapted to drinking salt water, the glands need to secrete at less than 20% of their maximum secretory capacity to eliminate the excess salts.

It is difficult to apportion the daily loss of water between the cloacal and nasal routes because the experiments that have been done rely on indirect calculations of water balance or on the short-term response to an administered salt load. The results obtained from short-term studies (lasting a few hours rather than days) are less meaningful because the amount of salt given to induce salt gland secretion is usually greater than would be ingested normally and exceeds that with which a bird could cope for an extended period, and the birds may enter a state of negative water balance (Table VI).

Ideally, the distribution of water loss should be studied over long periods (days) at steady-state, by direct measurements of salt gland secretion, cloacal and evaporative losses. This has not been done yet, but Holmes et al. (1968) and Fletcher and Holmes (1968) have estimated the distribution in domestic ducks given salt water to drink for periods greater than 14 days, by the use of simultaneous equations and the measurement or assumption of some of the variables (Table VII). Their figures indicate that in normally-fed ducks approximately 20% of the water loss occurs through the nasal glands, and in birds fasted for 24 h (when water consumption decreased), nasal loss goes to about 40%.

TABLE VI

Partition of electrolyte elimination between cloacal excretion and nasal secretion in the cormorant, *Phalacrocorax auritus*[a]

	Na (mEq)	%[b]	K (mEq)	%[b]	Cl (mEq)	%[b]	Water (ml)	%[b]
Total given	54	—	4.0	—	54	—	50	—
Cloacal excretion	25.6	52	2.66	90	27.5	51	108.9	68
Nasal excretion	23.8	48	0.31	10	26.1	49	51.4	32
Total excretion	49.4	—	2.97	—	53.5	—	160.3	—

[a] Sea water was given by stomach tube and the collections were for 8 h periods.
[b] Percentage of total excretion.
Modified from Schmidt-Nielsen *et al.* (1958) by Peaker and Linzell (1975).

Such data are completely lacking for truly marine species which eat invertebrates with a very high salt content. It is hoped that such studies will be done in order to apportion not only the losses of water, but also of ions and nitrogenous excretory products in birds exposed to extreme environmental conditions.

E. WATER BUDGETS

In order to see the relationships between the various pathways involved in water balance, and to distinguish adaptive patterns of control of water balance in relation to mesic, xeric, estuarine and marine habitats, one needs to compare complete water budgets of various representative species from each type of habitat. This is not yet possible, for with the

TABLE VII

Water budget of domestic ducks (*Anas platyrhynchos*) given hypertonic saline (284 mM Na, 6 mM K) to drink for periods exceeding 14 days[a]

Fed[b]		
ml kg body wt^{-1} day^{-1}		
Intake	Output	
180	Evaporative	16·8
	Cloacal	127·0
	Nasal	36·3
Fasted for 24 h[c]		
ml kg body wt^{-1} day^{-1}		
Intake	Output	
111	Evaporative	16·8
	Cloacal	49·3
	Nasal	45·0

[a] Calculated on the assumption that intake and output were at steady (i.e. isorrhoeic) state from

$$Z = W + X + Y \quad (1)$$
$$(Z \cdot Z_{Na}) = (W \cdot W_{Na}) + (X \cdot X_{Na}) + (Y \cdot Y_{Na}) \quad (2)$$

where W, X, Y and Z are evaporative water loss, rate of cloacal water loss, rate of salt gland secretion, and rate of water intake, respectively, in ml kg body wt^{-1} day^{-1} and W_{Na}, X_{Na}, Y_{Na} and Z_{Na} are the concentrations of sodium in these fluids, in mmol ml^{-1}. In both conditions, respiratory loss was assumed to be similar to that in the chicken measured by Barrott and Pringle (1941, cited by Holmes *et al.*, 1968).
[b] Measured variables are Z, Z_{Na}, X_{Na}, Y_{Na}.
[c] Measured variables are X, X_{Na}, Z_{Na}, Y_{Na}.
Fletcher and Holmes (1968); Holmes *et al.* (1968).

TABLE VIII

Water budgets of a few seed-eating birds in captivity that lack nasal salt glands[a]

Species, body weight (g), habitat type[b]	Treatment	Intake, ml kg^{-1} day^{-1}, % of total		Output, ml kg^{-1} day^{-1}, % of total			Reference
Zebra finch, *Poephila guttata*, 12·0, X	Eating seeds, deprived of water, $T_A = 24$, C, R.H. = 34%	preformed metabolic total	23·4 91·5 114·9	20·3 79·7	evaporation cloacal total	66·6 33·4 100·0[c]	66·6 33·4	Lee and Schmidt-Nielsen (1971)
Zebra finch, 13·4	Eating seeds, drinking fresh water, $T_A = 23$ C, R.H. = 70–90%	preformed metabolic drink total	43·3 149 284 476·3	9·1 31·3 59·6	evaporation[d] cloacal total	297 179·3 476·3	62·4 37·6	Skadhauge and Bradshaw (1974)
	As above, but deprived of water	preformed metabolic total	34·3 117 151·3	22·7 77·3	evaporation[d] cloacal total	91·7 59·6 151·3	60·7 39·3	Skadhauge and Bradshaw (1974)
Grey-backed finch-lark, *Eremopteryx verticalis*, 17, X	Eating seeds, assumed to be drinking the minimal requirement, $T_A = 24$–30 C, R.H. = 8%[e]	preformed metabolic drink total	11·8 63 32·2 107·1	11 59 30	evaporation cloacal total	89·5 17·6 107·1	84 16	Willoughby (1968)
Sage sparrow, *Amphispiza belli*, 18, X	Eating insects, assumed to be drinking minimal requirement, $T_A = 20$ C, R.H. = 16%[e]	preformed metabolic drink total	144 72·3 16·7 233	62 31 7	evaporation cloacal total	133 100 233	57 43	Moldenhauer and Wiens (1970)
Budgerigar, *Melopsittacus undulatus*, 30, X	Eating seeds, deprived of water, $T_A = 20$–25 C,	preformed metabolic total	13·3 48 61·3	22 78	evaporation cloacal total	39·7 21·6 61·3	65 36	Cade and Dybas (1962)

Species	Conditions	Input			Output			Reference
	As above, but T_A = 21–24 C, R.H. = 45–55%	preformed metabolic total	17.3 76.7 94	18 82	evaporation urine faeces total	62.0 30.3 1.7[a] 94	66 32 2	Krag and Skadhauge (1972)
Japanese quail, *Coturnix coturnix*, 105, M	Eating turkey starter mash, drinking fresh water *ad libitum*, T_A 23 C	preformed } metabolic drink total	12 208 220	5.5 94.5	evaporation[f] cloacal total	62.5 157.5 220	28 72	Chapman and McFarland (1971)
Bobwhite, *Colinus virginianus*, 180.5, M	Eating commercial turkey feed, drinking fresh water *ad libitum*, T_A = 20.5–28.0 C	preformed metabolic drink total	4 15 77 96	4 16 80	evaporation cloacal	31 32[g]	— —	McNabb (1969b)
Gambel's quail, *Lophortyx gambelii*, 148.8, X	As for Bobwhite	preformed metabolic drink total	4 17 75 96	4 18 78	evaporation cloacal	32 41[g]	— —	McNabb (1969b)
Chicken, *Gallus gallus*, 3150, M	Eating capon finishing pellets, drinking fresh water *ad libitum*, T_A = 20 C	preformed metabolic drink total	2.9 9.7 30.4 43	6.7 22.6 70.7	evaporation cloacal total	21.9 21.1 43	51 49	Dicker and Hasam (1972)

[a] The values have been calculated from data contained in the references.
[b] The letter X means the species is xerophilic, M means it is mesophilic.
[c] Note that intake exceeds output slightly. This explains how zebra finches can regain weight lost previously to dehydration, as observed by Lee and Schmidt-Nielsen (1971).
[d] Value inferred as the difference after subtracting the other output values from intake.
[e] This is the relative humidity at which evaporation rate was determined. At higher humidities, evaporation rate would be lower, and drinking would be correspondingly lower.
[f] Estimated from body weight by the equation of Crawford and Lasiewski (1968) (see Fig. 2).
[g] This is undoubtedly an underestimate, and the data as presented by McNabb do not make a balanced water budget.

exception of a few small desert seed-eaters, which show a remarkable ability to survive water deprivation, a few galliform birds and the domestic duck, data are insufficient to permit construction of a reasonable water budget. Furthermore, the water budget information that is available is not standardized as to experimental procedures, thus making comparisons still more difficult.

Table VII presents a water budget calculated for the domestic duck when fed and unfed and drinking hypertonic saline (simulating a wild duck in an estuarine habitat), and represents a bird that has salt-secreting nasal glands that play an important role in the water balance. Table VIII presents water budgets of several seed-eating birds that lack nasal salt glands. In ducks, as much as 40% of daily water turnover may involve the salt glands. In the Grey-backed finch-lark and budgerigar, both small, seed-eating desert birds well adapted to withstand dehydrating conditions, as demonstrated by the ability of individuals to survive many weeks without drinking on a diet of only dry seeds, from 66 to 84% of water loss is by evaporation, and the remainder is through the cloaca as urine and faeces. The comparison of water budgets of the three smallest desert birds show striking differences. The water turnover rate of the sage sparrow of North America was twice that of the Grey-backed finch-lark from southern Africa. Part of the sparrow's higher water turnover can be attributed to the insect food which would require more Na to be excreted in the urine than the seed diet of the lark. However, the sparrow's evaporative losses are greater, and unlike the lark, the sparrow could not survive water deprivation on a dry diet longer than an average of eight days (Moldenhauer and Wiens, 1970), and in this respect the sage sparrow is more like mesophilic birds than the other small xerophilic species shown. The smaller zebra finch when drinking water has a water turnover twice that of the sparrow; but when deprived of water its turnover is 50% less than that of the sparrow.

The data for the three species of quail, one a semi-domesticated Eurasian species, the others from North America, show how different the water turnover can be between species of similar size and habits. Although the smaller size of the Japanese quail will lead to a larger rate of water turnover, it seems unlikely that size alone could account for the two-fold difference between it and the species that are about 50% larger. Gambel's quail, native to the arid south-western United States and northern Mexico, does not show any difference in water budget from the Bobwhite, a native of more mesic habitats throughout North America, at least under the conditions of the measurements.

One way that any adaptations to aridity are likely to be revealed in the water metabolism is to determine the water budgets of birds restricted to the minimum amount of drinking water necessary to maintain water balance. Bartholomew and Cade (1963) have discussed this aspect and summarized data on *ad libitum* drinking versus minimal drinking requirements, showing that the *ad libitum* water consumption of a captive bird can be many times what it just needs to remain in water balance. When the Bobwhite and Gambel's quail were tested in this way, it turned out that the Gambel's quail is capable of a lower water metabolism, with a minimal drinking rate less than half that of the Bobwhite (McNabb, 1969a). This is consistent with the ability of Gambel's quail to produce urine with a markedly higher U/P ratio (Table V).

III. PATHWAYS AND CONTROL OF IONIC BALANCE

Integral with the maintenance of osmotic balance is the handling of ions by the renal-cloacal and nasal salt gland systems. In this section we discuss the structure and function of the principal organs of excretion in light of discoveries of the past ten years, and relate these functions to the osmotic and ionic balance of the organism.

A. RENAL STRUCTURE AND FUNCTION

1. *Organization of the Kidney*

The avian kidney has a complicated lobular organization that has made difficult the analysis of the details of its structure and function. However, recent anatomical studies have greatly illuminated the relationships between its structure and excretory processes, particularly in relation to production of hyperosmotic urine (O. W. Johnson, 1968, 1974; Johnson and Mugaas, 1970a, 1970b; Johnson *et al.*, 1972; Johnson and Ohmart, 1973).

The usual shape of the kidney is oblong, flattened dorsoventrally, with two, three or four large divisions, or lobes. The ureter runs anteroposteriorly along the ventral surface of the kidney, receiving urine discharged from branches that penetrate the kidney. The ends of the ureteral branches expand into clusters of medullary cones whose bases contact the lobular cortical tissue (Fig. 3). There are both arterial and renal portal blood supplies (Sperber, 1960). The cortical tissue is arranged in many cylindrical lobules, each having as its axis a branch of the renal vein called a central vein. The vein collects blood from a capillary plexus which receives blood from branches of the renal portal venous system called interlobular veins that are arranged around the periphery of the lobule, and from efferent glomerular arterioles. The nephrons are intertwined with this capillary plexus.

The nephrons discharge into collecting ducts that are situated peripherally on the cortical lobule. These collecting ducts then enter the medullary cones and converge toward the apex of the cone, eventually forming a secondary ureteral branch. There are two classes of nephrons, called reptilian type and mammalian type. The former, like the nephron

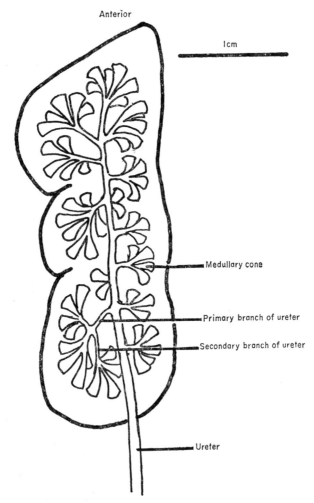

Fig. 3. Outline of a generalized avian left kidney, ventral view, showing the arrangement of ureteral branches and medullary cones.

of reptiles, lacks a loop of Henle, while the latter has a loop of Henle, like the mammalian nephron. These were excellently described by Braun and Dantzler (1972). The loops of Henle of the mammalian type nephrons extend into the medullary cones parallel with the collecting ducts. A single cortical lobule typically contributes collecting ducts and loops of Henle to several medullary cones, and each cone may be associated with more than one cortical lobule (Johnson et al., 1972). These relationships are illustrated in Fig. 4.

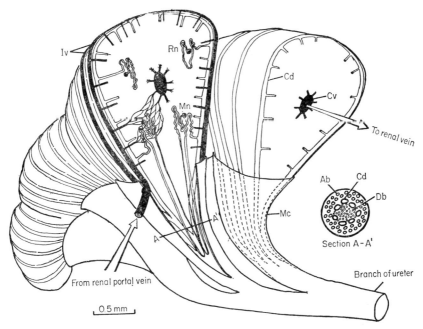

Fig. 4. The organization of medullary and cortical tissue in the avian kidney. Two adjacent cortical lobules are shown in section, each with a central (efferent) vein, and surrounded by collecting ducts arranged something like barrel hoops, but extending down into the medullary tissue and connecting with a branch of the ureter. On the left are seen parts of two other cortical lobules that contact each other and send collecting ducts into some of the same medullary cones. The sectioned lobule on the left shows how the mammalian-type and reptilian-type nephrons are related to the collecting ducts and the medullary cones. The arrangement of interlobular veins is indicated; parts of only three are shown but are otherwise omitted from the diagram. The diagram of section A-A' shows the arrangement of collecting ducts, and the descending and ascending arms of the loops of Henle. In many species, particularly of passeriforms, the thin-walled descending arms of the loops are clustered in the centre, surrounded by the large, thick-walled collecting ducts, which in turn are surrounded by the ascending arms, with walls somewhat thicker than in the descending arms. Blood capillaries permeate the cortical lobule and medullary tissue in the manner indicated. The arterial blood supply is omitted entirely (based on information in Braun and Dantzler, 1972; Johnson et al., 1972; Johnson and Mugaas, 1970a, 1970b; Johnson, 1974). Cd, collecting duct (= collecting tubule); Cv, central vein; Iv, interlobular vein; Mc, medullary cone; Mn, mammalian-type nephron; Rn reptilian-type nephron.

2. Countercurrent Multiplier

The parallel arrangement of the long loops of Henle and collecting ducts in the medullary lobules is analogous to the similar arrangement in the renal medulla of mammals, which is the anatomical basis for a large renal resorption of water and consequent production of hyperosmotic urine (Gordon, 1972). The observation that urine concentrating ability in some birds appears to be linearly related to the number of medullary lobules per unit volume, and hence also to the number of loops of Henle, led Poulson (1965) to hypothesize a countercurrent multiplier system of urine concentration like that of mammals. Descriptions of concentration gradients in renal medullary tissue fluid that increase from cortex to ureter in a variety of birds also supports this hypothesis (Skadhauge and Schmidt-Nielsen, 1967; Emery et al., 1972). Furthermore, NaCl is the principal solute involved in this concentration gradient, indicating that the countercurrent system works with a Na pump.

An examination of the relationship between medullary organization and urine concentrating ability indicates that there is a positive correlation between relative numbers of medullary cones and urine concentration (Johnson and Mugaas, 1970; Johnson, 1974). Some of this information is summarized in Table IX. It is clear that an increased proportion of medullary units could be correlated with a larger proportion of mammalian type nephrons, and hence a greater ability to concentrate the urine. Increased relative length of medullary units is not so clearly correlated with urine-concentrating ability as is their relative numbers.

3. Excretion

As in other vertebrates, the avian nephron produces urine by a process of filtration at the renal corpuscle, followed by selective resorption from and secretion into, the filtrate of various solutes as the fluid travels along the tubule. Water can be resorbed osmotically in the distal part of the tubule by active uptake of solutes such as Na^+ into the blood, and in the collecting duct as the urine passes through the increasing medullary concentration gradient on its way to the ureter. Adjustments of these processes of filtration, secretion and resorption produce urine at varying rates and of varying ionic composition and osmotic concentration, according to the organism's physiological requirements for maintaining homeostasis. Sperber (1960), Sturkie (1965) and Shoemaker (1972) have reviewed the literature on these processes.

a. *Glomerular filtration.* The glomerular filtration rate (GFR) increases with water loading, and decreases with dehydration or salt loading, as summarized in Table X. Reduction of GFR in hydrated birds occurs when arginine vasotocin, the antidiuretic hormone of birds, reptiles and amphibians, is administered intravenously (Skadhauge, 1964; Ames et al., 1971; Braun and Dantzler, 1974). These changes in GFR result mainly from changes in the number of filtering nephrons, rather than from changes in rate of filtration at individual nephrons; and, as shown in experiments with Gambel's quail, it is

TABLE IX

The relationship between relative length and numbers of renal medullary cones and urine concentrating ability

Species	Maximum U/P[a]	Relative length of medullary cones[b]	Mean kidney volume (mm³)[b]	Medullary cones (mm⁻³)[c]
Emu, *Dromaius novaehollandiae*	1·36	1·10	55000	—
Kookaburra, *Dacelo gigas*	2·71	3·00	1360	—
Senegal dove *Streptopelia senegalensis*	1·74	2·62	320	0·21–0·24
Mourning dove, *Zenaida macroura*	1·33	2·81	360	—
Budgerigar, *Melopsittacus undulatus*	2·33	5·06	125	0·48–0·50
Singing honey-eater, *Meliphaga virescens*	2·41	3·56	140	0·39–0·45
Zebra finch, *Poephila guttata*	2·78	4·71	47	0·85–0·96
House finch, *Carpodacus mexicanus*	2·36	3·59	126	0·44–0·50
Savannah sparrow, *Passerculus sandwichensis beldingi*	4·5	3·24	140	1·29–1·39

[a] Values from Table V.

[b] From Johnson (1974) and Johnson and Skadhauge (1975). Relative length = $\frac{\text{mean length} \times 10}{\text{cube root of kidney volume}}$.

[c] Calculated from number of medullary lobules per kidney in Johnson and Mugaas (1970). Note: the terms medullary cone and medullary lobule are synonymous.

TABLE X

Glomerular filtration rates observed in several studies

Species	Treatment	GFR ml kg^{-1} min^{-1}	Reference
Chicken, *Gallus gallus*	Normally hydrated	1·23	Korr (1939)
	100 ml H$_2$O by stomach tube	2·47	Korr, (1939)
	Normally hydrated controls	1·23	Dantzler (1966)
	After intravenous infusion of 45–50 mEq NaCl kg^{-1} as 6% NaCl	0·49 ("40% of control value")	Dantzler (1966)
	Dehydrated	1·734 ± 0·314 (mean ± s.e.)	Skadhauge and Schmidt-Nielsen (1967b)
	Water by stomach tube, 9% of dehydrated body weight	2·123 ± 0·474	Skadhauge and Schmidt-Nielsen (1967b)
	Intravenous infusion of 18% NaCl	2·066 ± 0·552	Skadhauge and Schmidt-Nielsen (1967b)
Gambel's quail, *Lophortyx gambelii*	Hydrated control (hypoosmotic infusion)	0·882 ± 0·036	Braun and Dantzler (1972)
	After intravenous infusion of 40 mEq of NaCl kg^{-1} as 6% NaCl	0·176 ("20% of control value")	Braun and Dantzler (1972)

reptilian-type nephrons, not mammalian type, that are involved in these changes (Braun and Dantzler, 1972, 1974). Arginine vasotocin appears to act selectively on the glomerular arterioles of the reptilian-type nephrons.

It would appear then that during diuresis, most or all of the nephrons are filtering. During salt loading or dehydration, arginine vasotocin released by the neurohypophysis acts to reduce or stop filtration in reptilian-type nephrons, leaving the mammalian-type nephrons with their loops of Henle to continue to filter, and through the counter-current multiplier, to promote resorption of water in the collecting ducts in order to enhance the concentration of the urine.

b. *Excretion of Na^+, K^+ and Cl^-.* As is to be expected, the concentrations of these monovalent ions in urine vary according to the rates of intake with food and drink, and with the need to conserve water by concentrating the urine. Birds that eat meat tend to excrete higher concentrations of Na than those that eat seeds or other plant material, and the latter excrete more K, owing to the different relative amounts of these ions in the different foods. Seeds in particular are low in Na and high in K. Also, birds that drink sea water or brackish water must excrete the excess NaCl at least partially in the urine. Table XI presents concentrations of these ions in the urine of several species of birds under different conditions of feeding and drinking. These values do not include ions bound with the undissolved components of the urine, such as urates (see Section II.B.1).

The variations in concentrations of these ions stem mostly or entirely from their differential resorption in the nephron. In chickens, for example, from dehydration to water loading, urinary Na excretion increased from 1·52 to 2·40% of the filtered load, and Cl excretion increased from 0·84 to 2·36% of the filtered load; and at the same time, K excretion decreased from 16·2 to 13·3% of the filtered load, although the urine volume increased by a factor of 17 (Skadhauge and Schmidt-Nielsen, 1967b). There is some evidence that K is sometimes secreted into the tubule, as well (Sperber, 1960).

c. *Excretion of divalent ions.* Much less is known about excretion of the divalent ions such as Ca^{2+}, Mg^{2+} and SO_4^{2-} than about monovalent ions. One of the reasons for this is that these ions are often more difficult to analyse quantitatively than are the monovalent ions, a difficulty that has to some degree discouraged workers from studying them in birds, which are small and produce scanty urine. An understanding of how birds handle these ions is of interest to comparative physiologists because these divalent ions occur in appreciable amounts in sea water and in "alkaline" desert springs and pans, so birds that must depend on such sources of water have to deal with these ions.

TABLE XI

Urinary concentrations of sodium, potassium and chloride in various birds under different conditions of feeding and drinking

Species	Feeding and drinking conditions	mEq l⁻¹, mean ± s.e.			Reference
		Na	K	Cl	
Birds eating foods derived from plants					
Singing honeyeater, *Meliphaga virescens*	Eating honey and wheat germ, drinking fresh water	23·4 ± 6·4	130 ± 11	24·1 ± 3·1	Skadhauge (1974b)
	As above, deprived of water	11·8 ± 1·0	114 ± 8	24·3 ± 1·0	Skadhauge (1974b)
Zebra finch, *Poephila guttata*	Eating seeds, drinking fresh water	7·5 ± 0·7	42 ± 4·6	19·3 ± 1·2	Skadhauge (1974b)
	Eating seeds, deprived of water	7·6 ± 0·6	135 ± 5	39·8 ± 3·6	Skadhauge (1974b)
Budgerigar, *Melopsittacus undulatus*	Eating seeds, drinking fresh water	6·3 ± 4·5	41·7 ± 18·3	22·0 ± 14·7	Skadhauge (1974b)
Crested pigeon, *Ocyphaps lophotes*	Eating seeds, drinking fresh water	8·8 ± 1·5	151 ± 19	17·4 ± 2·6	Skadhauge (1974b)
	Eating seeds deprived of water	14·0 ± 1·5	162 ± 13	23·0 ± 6·5	Skadhauge (1974b)
Turkey, *Meleagris gallopavo*	Eating poultry feed, deprived of water	81·0	105	97·0	Skadhauge and Schmidt-Nielsen (1967b)
	Water-loaded by intravenous infusion	5·5	9·2	2·5	Skadhauge and Schmidt-Nielsen (1967b)
	Salt-loaded by intravenous infusion	227	77·3	248	Skadhauge and Schmidt-Nielsen (1967b)

Species	Condition				Reference
Emu, *Dromaius novae-hollandiae*	Eating wheat and poultry feed, drinking fresh water	6.0 ± 1.4	77 ± 6	16.2 ± 3.3	Skadhauge (1974b)
	As above, deprived of water	6.1 ± 1.2	120 ± 11	4.7 ± 0.6	Skadhauge (1974b)
Duck, *Anas platyrhynchos*	Drinking fresh water	0.57 ± 0.29	1.96 ± 0.41	0.709 ± 0.29	Holmes et al. (1968)
	Drinking hyperosmotic saline, (284 mmol NaCl, 6.0 mmol KCl)	4.37 ± 0.86	2.18 ± 0.37	6.70 ± 1.23	Holmes et al. (1968)
Birds eating meat[a]					
Kookaburra, *Dacelo gigas*	Eating meat, drinking fresh water	38.3 ± 4.6	58.4 ± 15.2	33.2 ± 1.4	Skadhauge (1974b)
	Neither drinking nor eating	28	92.5	9.9	Skadhauge (1974b)
Roadrunner, *Geococcyx californianus*	Eating meat, drinking fresh water	70 ± 14	43 ± 4	—	Calder and Bentley (1967)
	Neither drinking nor eating	70 ± 9	75 ± 29	—	Calder and Bentley (1967)
Pelican, *Pelecanus erythrorhynchos*	Eating fish, drinking fresh water	35 ± 10	16 ± 4	—	Calder and Bentley (1967)
	Neither drinking nor eating	18 ± 6	114 ± 26	—	Calder and Bentley (1967)
	Eating fish, not drinking	47 ± 22	39 ± 4	—	Calder and Bentley (1967)
Bateleur eagle, *Terathopius ecaudatus*	Eating meat	45	77	40	Cade and Greenwald (1966)
Gabar goshawk, *Micronisus gabar*	Eating meat	73	48	48	Cade and Greenwald (1966)
American kestrel, *Falco sparverius*	Eating meat	103	41	89	Cade and Greenwald (1966)

[a] Except for the Kookaburra, all these species have nasal salt glands.

TABLE XII

Excretion of inorganic ions by ducks

Treatment	Urine concentration, mM, mean ± s.e.				Renal excretory rate, mmol kg⁻¹ day⁻¹, mean ± s.e.			
	Na^+	K^+	Ca^{2+}	Cl^-	Na^+	K^+	Ca^{2+}	Cl^-
Drinking fresh water	0·49 ± 3·7	37·0 ± 6·3	1·25 ± 0·19	11·6 ± 3·76	0·57 ± 0·29	1·96 ± 0·41	0·0687 ± 0·017	0·709 ± 0·29
Drinking hyperosmotic saline (284 mM NaCl, 6·0 mM KCl)	76·3 ± 2·9	45·5 ± 6·9	3·93 ± 0·71	124 ± 12·9	4·37 ± 0·86	2·18 ± 0·37	0·211 ± 0·040	6·70 ± 1·23

Data from Holmes et al. (1968).

Holmes et al. (1968) examined excretion of Ca^{2+} and other ions in ducks drinking fresh water and hypertonic saline solution. They observed an increase in the urinary excretion of Ca in the saline-drinking ducks (Table XII), which also had increased excretion of NaCl.

Willoughby (1971) studied excretion of Ca and Mg in Japanese quail that were given strong solutions of $CaCl_2$ and $MgCl_2$ by mouth (drinking or by stomach tube). Urine was collected in some cases as birds expelled it spontaneously (i.e. as cloacal "urine") and in other

TABLE XIII

Concentrations of cations in urine of quail (*Coturnix coturnix*) after drinking solutions of $CaCl_2$ and $MgCl_2$ for several days

Regimen	mEq l^{-1}, mean ± 2 s.e.			
	Ca^{2+}	Mg^{2+}	Na^+	K^+
Cloacal urine				
Distilled water (control)	8·7 ± 3·9	4·8 ± 2·0	9·3 ± 3·6	35·2 ± 12·5
0·068 M $MgCl_2$ (136 mEq l^{-1} Mg)	21·3 ± 10·3	121·5 ± 30·4	41·5 ± 14·9	17·0 ± 4·9
0·084 M $MgCl_2$ (168 mEq l^{-1} Mg)	52·0 ± 34·4	128·0 ± 53·9	39·6 ± 14·7	24·6 ± 8·3
0·072 M $CaCl_2$ (144 mEq l^{-1} Ca)	125·4 ± 61·7	15·2 ± 6·8	36·7 ± 37·0	41·5 ± 13·9
0·090 M $CaCl_2$ (180 mEq l^{-1} Ca)	187·6 ± 50·0	20·3 ± 5·7	37·1 ± 17·7	44·9 ± 10·0
Ureteral urine				
Distilled water (control)	6·1 ± 2·9	18·9 ± 5·7	14·4 ± 5·9	91·5 ± 16·0
0·068 M $MgCl_2$	5·9 ± 9·5	46·3 ± 19·9	18·0 ± 15·5	125·0 ± 123·5
0·072 M $CaCl_2$	3·4 ± 1·5	17·1 ± 3·9	9·0 ± 3·8	49·2 ± 15·7

Willoughby (1971).

cases by inserting a collecting tube into the cloaca to obtain ureteral urine. These samples were treated with nitric acid to dissolve cations that may have been present in undissolved compounds such as urates and carbonates.

Cloacal urine of quail receiving 0·084 M $MgCl_2$ for several days contained elevated concentrations of Mg as expected, but also had concentrations of Ca averaging six times that of control birds receiving distilled water. Cloacal urine of quail receiving 0·072 M or 0·090 M $CaCl_2$ for several days had concentrations of Ca elevated to match the concentrations in the solutions, but they also had elevated concentrations of Mg, compared to controls. Furthermore, plasma of quail receiving chronic doses of $MgCl_2$ had slightly reduced concentrations of Ca and slightly elevated concentrations of Mg. When quail received either $CaCl_2$ or $MgCl_2$ solutions, cloacal urine had significantly elevated amounts of Na.

On the other hand, ureteral urine of these quail was different. A load of $MgCl_2$ resulted in elevated urinary Mg, but not of Ca or Na. Concentrations of K in urine were unaffected by any treatment, except that ureteral urine contained more K than did cloacal urine. Table XIII summarizes some of these observations.

These results indicate that Mg^{2+} and Ca^{2+} interact with a common regulatory pathway in the body. The pathway may involve a common transport system in the gut or in the nephron or both, and probably involves the actions of parathormone. The interaction of Na^+ with the divalent cations is also intriguing, and needs further study.

B. POST-RENAL MODIFICATION OF URINE

It is now well established that active ion movements can occur across the walls of the rectum and cloaca and thus modify the ionic composition of ureteral urine. In studies involving luminal perfusion of the rectum and cloaca of anaesthetized chickens, Skadhauge and his associates (Skadhauge, 1973) have shown that Na can be actively resorbed and K secreted. The absorption of Na and Cl appears to be under systemic control, since absorption of these ions was decreased considerably after the administration of NaCl to the birds. However, the factors responsible for the control are unknown. Also involved are local control mechanisms. For example, a high urine K concentration decreases Na absorption, and Na movements are coupled to those of K to some extent.

While these studies demonstrate that ion movements can occur, it is difficult to assess their significance in the overall ionic balance of a bird and even more difficult to assess their role in birds in different physiological states or in those adapted to different habitats and diets, because factors other than the transport mechanisms are involved. Thus a knowledge of the extent to which urine flows into the rectum, the time it is stored there, and whether it is mixed with faeces is needed to obtain an overall view. Unfortunately, this information is the most difficult to obtain. Nevertheless, there are important differences between species in respect to the properties of the transport mechanisms, and Skadhauge (1974a) has found that in perfusion experiments in the galah, a high urine K concentration does not inhibit Na absorption as in the chicken. Therefore, during dehydration, when urine K increases markedly, Na absorption can continue. This may be an important adaptation in birds with a low dietary intake of Na, and with the difference in the solute-linked water flow (Section II. B) the galah can conserve 70% of the urine Na without losing water to the urine in the cloaca, which is hypertonic to the plasma. Probably also in the chicken, Na conservation is important, and it has long been known that if ureteral urine is diverted to the exterior by surgical means, these birds become deficient in Na (Hart and Essex, 1942).

Another important point has been raised by Skadhauge (1975).

> ... the major osmolytes in the urine of dehydrated birds are not strong electrolytes and an urgent problem for future work in this field is what constitutes the rest of the osmotic concentration in such birds. Further questions raised, of course, are whether these other solutes are resorbed in the cloaca and whether the solute movement gives rise to solute-linked water flow. These questions have perhaps been overlooked previously, although there is some justification for solving problems of sodium, potassium and chloride movements first. But the time is now ripe for other studies and, in particular, ammonium compounds should be investigated since ammonia may constitute an important fraction of urine osmolality in dehydrated birds. (Skadhauge, 1975.)

Schmidt-Nielsen *et al.* (1963) suggested that salt gland secretion may be required to take advantage of any water resorption in the cloaca. They argued that if water movement is linked to that of Na, then Na would be present in the body in excess and extrarenal excretion of NaCl at high concentrations would be necessary for the efficient resorption of water in the cloaca. However, it is now known that birds without salt glands such as the chicken reabsorb Na cloacally for Na conservation, so a correlation between salt gland and cloacal Na resorption does not hold in all species. Nevertheless, an important topic for future research is cloacal transport in marine birds, to see whether they can resorb more Na and water cloacally than other birds. There is some evidence for Na resorption in the cloaca of marine birds (Peaker and Linzell, 1975). However, Holmes *et al.* (1968) have suggested that the shorter the time when hypertonic urine of salt-adapted birds is in contact with the cloacal walls the less the chance it would have of being diluted by the osmotic passage of water from the blood, and they found that voiding occurred more frequently in ducks kept on salt water. This form of behavioural control might be an important means of determining the extent to which urine is altered in the cloaca. However, until it is known whether water and Na resorption could be an advantage or, as they suggest, it would be a disadvantage, one can only speculate.

C. NASAL SALT GLANDS

Since Schmidt-Nielsen *et al.* (1957, 1958) demonstrated that the nasal glands of marine birds secrete large amounts of hypertonic NaCl solutions these organs have been of considerable interest because they represent an extreme adaptation of cellular processes in their ability

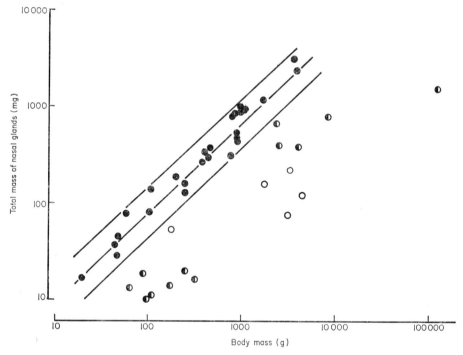

Fig. 5. Relationship between body weight and nasal gland weight in different birds; ●, truly marine birds; ○, birds habitually drinking fresh water (those tested incapable of secreting salt); ◐, terrestrial and brackish-water birds capable of secreting. The regression line and 95% confidence limits are for the marine birds ($y = 0.09 + x^{0.92}$, where y = weight of glands (mg) and x = body weight (g) (modified from Peaker and Linzell, 1975).

to produce such concentrated and copious solutions. We do not have space here to give an extensive account, so the more recent studies on control of secretion and on ion transport are stressed. Readers seeking more information are referred to the recent monograph by Peaker and Linzell (1975).

The glands are larger in marine species than in other birds of similar size (Fig. 5). However, relatively few studies have involved truly marine species, and most experiments are done with gulls or the domestic duck (*Anas platyrhynchos*) and goose (*Anser anser*). The salt

TABLE XIV

Orders of birds in which salt gland secretion has been observed

---
Struthioniformes
Podicipediformes
Sphenisciformes
Procellariiformes
Pelecaniformes
Anseriformes
Phoenicopteriformes
Ciconiiformes
Falconiformes
Galliformes
Gruiformes
Charadriiformes
Gaviiformes
Cuculiformes

---

glands of these species begin secreting within minutes of the administration of hypertonic saline solutions even if the birds have not previously been exposed to high salt intakes.

Besides marine birds, some terrestrial species have salt glands. Current evidence indicates their presence in 14 orders of living birds (Table XIV).

1. *Structure in Relation to Function*

The nasal glands are situated in or around the orbit but in most marine and estuarine birds they are located above the eye and receive a rich blood supply and innervation from below. The secretion is formed in secretory tubules which radiate from the central canals of the lobules; the central canals join to form secondary ducts which in turn enter the primary ducts. It is these primary ducts (usually two, sometimes one, on each side) which carry the fluid to the nasal cavity (Fig. 6) (Fänge *et al.* 1958a).

The secretory cells show signs of great metabolic activity: the basal membrane is greatly infolded and the cytoplasm is packed with mitochondria (Ernst and Ellis, 1969) (Fig. 7). The cells at the periphery of the lobules, i.e the end of the tubules distal to the central canal, are unspecialized and Ellis (1965) has concluded from labelling studies that they are the site of cell division in the tubule. In adult birds the

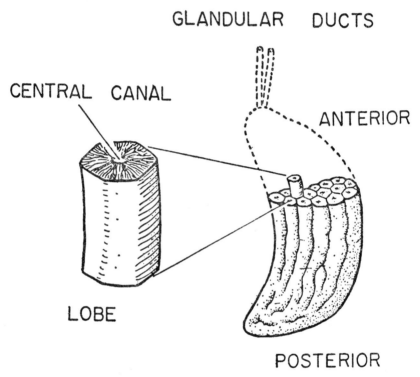

Fig. 6. The arrangement of lobes in salt gland of the gull, *Larus argentatus* (from Fänge et al., 1958a).

majority of the cells in the tubule are principal or secretory cells. There is a gradient of cell types along the tubule, those with more mitochondria and greater basal infolding being situated adjacent to the central canal (Ernst and Ellis, 1969). The degree of specialization of the cells and the length of the tubules are affected by the salinity of the drinking water.

The ultrastructure of the several layers of cells which line the duct system suggests that they are not highly active and it appears that they are probably not involved in modifying the primary secretion as in some exocrine glands but instead act as simple conduits with impermeable

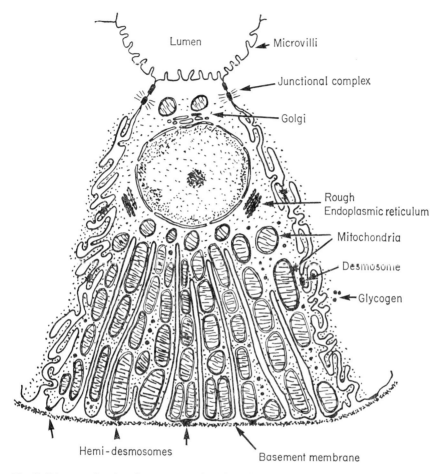

Fig. 7. Diagram showing the structure of a salt and gland secretory cell (from Peaker and Linzell, 1975).

walls to carry the secretion to the exterior, at least in marine birds (Peaker and Linzell, 1975).

The blood supply of the lobules is arranged in a manner such that the highly specialized secretory cells at the centre receive blood with the highest NaCl content from the arterial end of the capillaries; the capillary blood then runs in a countercurrent manner to the flow of secretion before entering a venous plexus at the periphery of the lobules (Fänge et al., 1958a; Peaker and Linzell, 1975). Thus blood passes the bases of many secretory cells—an arrangement which could help account for the high rates of extraction of Na, Cl, K and water from the blood, even with the very high rates of blood flow through the gland. Hanwell et al. (1971a) calculated extraction in geese, using the Fick principle, from the rates of blood flow and secretion; the median figures (expressed as a percentage of that arriving in arterial plasma secreted per unit time) were: Cl, 21%; Na, 15%; K, 35% and H_2O, 5·8%. However, in some geese with very high secretory rates, figures of up to 70% for Cl and 57% for Na were obtained. Incidentally, the rates of blood flow during secretion, 5–15 ml (g tissue min)$^{-1}$, are amongst the highest recorded for any organ and indicate the extreme activity displayed by these small glands.

2. Composition of the Secretion

The typical composition of nasal fluid from a marine bird is shown in Table XV. It is clear that the main constituents are Na and Cl in high

TABLE XV

Typical composition of nasal fluid in the gull, Larus argentatus

Solute	Concentration, mmol
Sodium	718
Potassium	24
Calcium and magnesium	1
Chloride	720
Bicarbonate	13
Sulphate	0·35

Modified from Schmidt-Nielsen (1960).

concentrations, and that these ions account for 97% of the total osmolality. Schmidt-Nielsen (1960) and Zaks and Sokolova (1961) showed that the NaCl concentration is higher in pelagic, invertebrate-eating species than in those that live on coasts or in estuaries, or that eat fish; in other words the concentration is higher in those birds that ingest more hypertonic drinking water and food. The maximum concentration recorded for a marine bird is 900–1100 mmol in the petrel Oceanodroma leucorrhoa.

Probably the maximum ability of the glands in each species to concentrate ions is characteristic of that species. Even so, many studies show that there can be an adaptation of the gland's secretory capacities, including its maximum concentrating ability, when individuals previously given fresh water to drink are given sea water, and vice versa.

Secretion of a similar composition to that of marine species is produced by some terrestrial birds—falconiforms (Cade and Greenwald, 1966), the cuculiform *Geococcyx californianus* (Ohmart, 1972) and the galliform *Ammoperdix heyi* (Schmidt-Nielsen et al., 1963). However, the nasal secretion of the ostrich (*Struthio camelus*) stimulated to secrete by high ambient temperatures and deprivation of drinking water, has high concentrations of K, Na, Ca and Cl. The composition of this nasal secretion was rather variable, and in some samples Na and K concentrations were similar, but in others K was 5–10 times more concentrated than Na (Schmidt-Nielsen et al., 1963).

3. Control of Secretion

Following the early studies of Fänge et al. (1958b) on the nervous control of secretion, it is now well established that secretion is induced and maintained by a secretory reflex, the efferent limb being cholinergic and acting to stimulate the cells of the gland and to cause vasodilation in the gland during secretion. However, the receptors which detect excess salt in the body have been found, by experiments involving infusion, cross-circulation, perfusion and denervation, to be located not in the brain or elsewhere in the head but in the region of the heart, with the afferent limb of the reflex running in the vagus nerves (Fig. 8) (Hanwell et al., 1972).

There has been some argument over what constitutes the primary stimulus, i.e. whether osmoreceptors, blood volume receptors or specific ion receptors are involved. Fänge et al. (1958b), who found that the intravenous administration of sucrose initiated secretion, postulated the action of osmoreceptors. However, since blood volume increases markedly after salt loading, as water is drawn osmotically from the

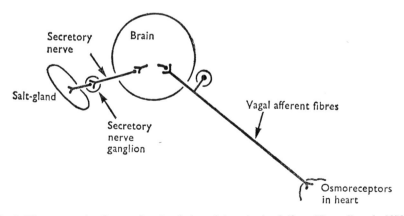

Fig. 8. The proposed reflex arc for stimulation of the salt gland (from Hanwell et al., 1972).

tissues, Donaldson and Holmes (Holmes, 1965) and Burford and Bond (1968) suggested that volume or stretch receptors might trigger secretion. Furthermore, apparent support for this hypothesis was obtained by Donaldson and Holmes, who found that the intravenous administration of isotonic saline (0·154 M NaCl) containing gum arabic initiated secretion, and by Hughes (1972) who observed secretion following the oral administration of 0·154 molar NaCl.

However, later experiments by Hanwell et al. (1971b, 1972) and by Stewart (1972) showed that blood volume may not be increased or may be decreased when secretion starts after oral salt loading or during a period of water deprivation. Moreover, Hanwell et al. (1971b) found in geese that plasma Na and Cl increased after giving 0·154 M NaCl, and that secretion was initiated; so it now appears highly likely that so-called isotonic saline is in fact hypertonic as far as the membranes of the receptors are concerned. Experiments with oral loads of relatively weak solutions, such as those done by Hughes (1972) are difficult to interpret because of the added complication of passage of water and solutes out of and into the alimentary canal. Direct evidence was obtained against the hypothesis that blood volume receptors are responsible by Hanwell et al. (1972) who rapidly injected homologous blood into the right atrium in volumes sufficient to raise the blood volume by 9–16%; secretion was not induced. Nevertheless, the injection of hypertonic NaCl soon after did stimulate secretion.

Several groups of workers have studied the effects on secretion of administering various solutions (Hanwell et al., 1972; Peaker and Linzell, 1975). The salt gland receptors have strikingly similar responses to those discovered by Verney (1947) in mammals which control release of antidiuretic hormone. The requirement for stimulation is not simply an increase in the solute content of plasma but in tonicity which involves net water movement out of a cell in response to an osmotic gradient. Thus, for tonicity to be increased, the cell membrane must be relatively impermeable to the solute. Therefore, hyperosmotic solutions of NaCl, sucrose, mannitol, lithium chloride, sodium sulphate, etc. induce secretion, whereas hyperosmotic solutions of urea and glucose do not. Current evidence therefore favours the view that osmoreceptors rather than specific ion receptors are involved. Unfortunately some authors still do not make the distinction between osmolality and tonicity, so Hanwell et al. (1972) suggested that tonicity receptors might be a better term to use.

An interesting difference between species, so far unexplained, was found by Peaker et al. (1973). The receptors in larger birds appear to be more sensitive than in smaller birds in that a smaller increase in plasma Na concentration is required to initiate secretion (Fig. 9).

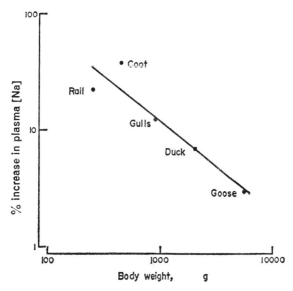

Fig. 9. Relationship between body weight and percentage increase in plasma sodium concentration required to initiate salt gland secretion. The slope of the log–log line is −0·762, and the correlation coefficient is −0·932 ($p < 0·05$) (from Peaker et al., 1973).

Although hormones may affect secretion, current evidence favours the view that they play only a permissive or secondary role, and control of secretion is primarily a function of the nervous reflex (Phillips and Ensor, 1972; Peaker and Linzell, 1975).

4. The Secretory Mechanism

Salt gland secretion is initiated and maintained by cholinergic nerves, and so one must not only consider how the secretory cells transport ions but also how acetylcholine stimulates the mechanism.

The scheme currently favoured for ion transport is shown in Fig. 10. It is based on measured concentrations and potential gradients, the effects of metabolic inhibitors and respiratory stimulants, and on transport of other ions (Peaker 1971; Peaker and Linzell, 1975). To produce a hypertonic secretion, the luminal membrane must have a low osmotic permeability to water, and a low passive permeability to ions. The basal infoldings seen in adult birds and especially in marine species might also constitute a backwards-facing standing-gradient system on the lines outlined by Diamond and Bossert (1968). However, the glands of young birds which lack such infoldings can still form concentrated solutions and so Peaker and Linzell (1975) have suggested that the concentration of the fluid is determined by the properties of the luminal membrane, but that a standing gradient operating in a manner such that a hypertonic solution enters the mouths of the channels may serve to prevent an imbalance in water movements into and out of the excretory cells, and that hypotonicity of the fluid in the channels at the closed end may act to ensure a steep concentration gradient for ions to pass down from blood, thus ensuring that the efficiency of extraction is high.

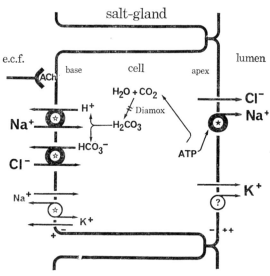

Fig. 10. Scheme for ion transport in the salt gland; open star: non-electrogenic pump closed star: electrogenic pump (from Peaker, 1971).

The manner in which acetylcholine stimulates activity is not known but the primary event may be stimulation of the Na pump on the luminal membrane (Peaker, 1971).

5. *Adaptive Patterns of Nasal Secretion*

a. *Marine and estuarine birds.* By secreting a NaCl solution more concentrated than that ingested, the glands act to supply osmotically free water so that the kidney can excrete nitrogenous wastes and K. The excess NaCl ingested is derived from sea water and from the food which, if the main diet is invertebrates, contains Na, Cl and other ions in relatively large amounts. Thus the glands play a vital role in such species.

Other features of marine and estuarine birds are: renal excretion is modified so that Na is reabsorbed (and excreted via the salt glands) so that Na occupies little of the "available osmotic space" of the urine;

intestinal absorption of Na and Cl is increased but magnesium and sulphate largely pass through the intestine unabsorbed; drinking behaviour appears to change in a manner such that birds do not drink sea water in excess of their ability to excrete the sodium thus obtained (Holmes and Wright, 1969; Peaker and Linzell, 1975).

The pattern of ionic balance in a marine bird is diagramatically depicted by Fig. 11. A similar pattern probably applies to birds which live by salt water either in coastal marshes (e.g. the crane *Grus rubicundus*, Hughes and Blackman, 1973) or in inland salt lakes and sloughs. In the latter regard, Cooch (1964) gives a lucid discussion of the

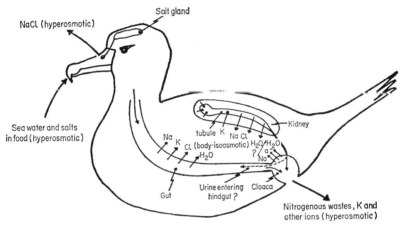

Fig. 11. Diagram showing the main routes of ion movements and excretion in a marine bird.

importance of the salt gland in the survival of wild ducks that visit salty (or alkaline) prairie ponds in North America.

An important feature of salt glands is their ability to adapt to the prevailing conditions of environmental salinity. In the wild this probably is particularly important in species that migrate between fresh water nesting grounds and estuarine wintering grounds. When a bird starts to ingest salt water, the glands increase in size, the secretory ability (as output per gland and per unit weight of gland) and the concentration of the fluid increase within a few days. There are concomitant changes in nucleic acids and enzymes in the gland tissue (Fig. 12) (Holmes, 1972). Furthermore, the tubules contain more highly specialized cells (Fig.13) (Ernst and Ellis, 1969; Komnick and Kniprath, 1970). The changes are reversed when birds are transferred to fresh water.

The trigger for adaptive hypertrophy of the glands appears to be nervous (Hanwell and Peaker, 1973, 1975) and Hanwell and Peaker (1975) have suggested that secretory activity and hypertrophy may be obligatorily linked.

b. *Terrestrial carnivorous birds*. A number of falconiforms secrete highly concentrated nasal fluid (0·4 to 2·4 M NaCl) (Cade and Greenwald,

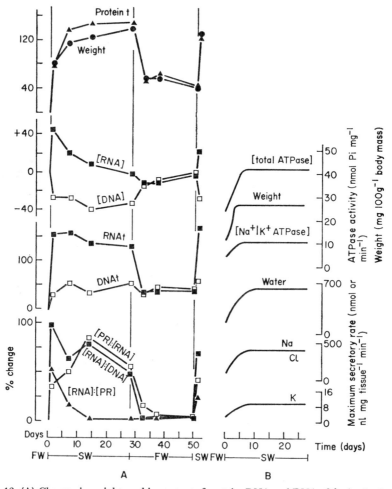

Fig. 12. (A) Changes in weight, and in content of protein, RNA and DNA of duck salt glands during adaptation to salt water, deadaptation and readaptation. Mean values are shown, expressed as a percentage change from the corresponding value for birds on fresh water (FW) (redrawn from Holmes and Stewart, 1968). (B) Changes in weight, maximum secretory ability and ATPases in duck salt glands during adaptation to salt water. The line which best describes the mean changes is shown, and the time scale has been drawn as in (A) (modified and redrawn from Fletcher et al., 1967). In both series of experiments the hypertonic drinking water (SW) contained 284 mM Na and 6 mM K. Concentrations are shown in square brackets, total gland content with subscript (from Peaker and Linzell, 1975).

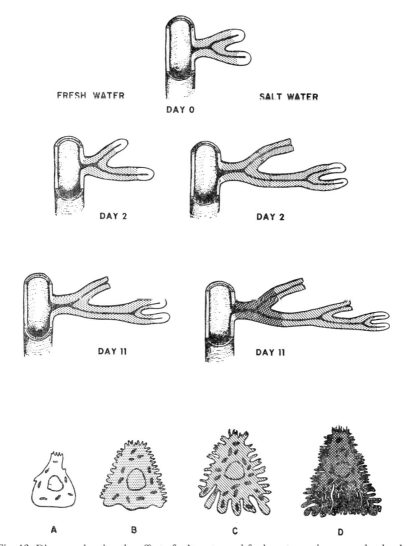

Fig. 13. Diagram showing the effect of salt water and fresh water regimens on the development of cellular specialization in the secretory epithelium of the salt gland. Peripheral cells (A) exhibit little specialization of their cell surfaces and contain few mitochondria. Partially specialized secretory cells (B) are characterized by short folds along their lateral surfaces and flat basal membranes; the distribution of mitochondria in this cell type is similar to that of the peripheral cells. Two stages in the development of the fully specialized secretory cell type (C and D) are shown. In the first stage (C) the cells exhibit some folding of the basal surface in addition to the lateral plications; mitochondria, more numerous in this cell type in the partially specialized secretory cell, are fairly evenly distributed in the cytoplasm. In the second stage (D) the specialized secretory cell is fully developed: both the lateral and basal membranes are extensively infolded, forming complex intracellular compartments and extracellular spaces: the mitochondria are increased dramatically in number and pack the basal labyrinths. The distribution of these cell types in fresh water and salt water-adapted glands on various days on the regimen is shown for a single secretory tubule as it grows out from the central canal. In addition to the development of cellular specialization, the tubules of birds given salt water elongate more rapidly and show more branching than birds kept on fresh water (from Ernst and Ellis, 1969).

1966) but the amounts involved are so small it is difficult to believe that the salt glands are important, and Johnson (1969) has shown that in *Buteo jamaicensis* the glands only account for about 3% of the Na excreted. However, Cade and Greenwald (1966) suggested that salt secretion may be more important in nestlings than in adults since their only source of water is the food brought by the adults and, in arid regions, evaporative water loss is likely to be high. Incidentally, such a situation might also occur in nestlings of marine species. Ohmart (1972) described nasal salt secretion in the roadrunner *Geococcyx californianus*, which lives in desertic regions, and concluded that the salt glands are of greatest survival value in nestlings. In the nest they are exposed to solar radiation and the salt glands appear to counterbalance water loss for evaporative cooling while the parents are gathering food and cannot provide shade. Ohmart does not believe that the salt glands are used to any great extent by free-ranging adults, and in fact the weight of the glands decreases about three-fold relative to body weight as the individual matures.

c. *Terrestrial omnivorous birds*. The only species for which there is any real information is the ostrich, in which the composition of the nasal secretion is labile, as noted above. The extent to which the glands play a role in the ionic and osmotic balance remains unknown.

IV. SUMMARY

The principal organs involved in osmotic and ionic regulation in birds are the gut, kidney and nasal salt gland. It is through the gut that water and solutes are taken into the body. Water and salts leave the body via urine and faeces, by evaporation (water loss only), and (in certain birds) by secretion from nasal salt gland. The few studies of water metabolism indicate that daily turnover rates are proportional approximately to body weight to the 0·7 power. Evaporation accounts for 30 to 84% of the turnover. Over 50% of evaporation occurs from the skin, the rest from respiratory surfaces, in six of seven species studied. Individuals of some species can reduce their total evaporation rate when adapting to dehydrating conditions.

Water loss through the kidney is minimized by producing hyperosmotic urine with maximum urine to plasma ratios ranging from 1·3 to 4·5, and by resorption of a certain amount of water from the urine while it is stored in the cloaca and rectum. Birds with active nasal salt glands, such as ducks drinking salt solutions, may lose up to 40% of total water loss through nasal salt secretion.

Water budgets of many more species are needed to clarify adaptive patterns in avian water metabolism.

The avian kidney is organized into a system of cortical and medullary lobules. Two classes of nephrons, the reptilian type which lacks a loop of Henle, and the mammalian type which has a loop of Henle, occur in the cortical lobules. The loops of Henle extend into the medullary lobules parallel with the collecting ducts in an arrangement that permits resorption of water by a countercurrent multiplier system based on the active transport of Na^+. The rate of urine production is varied partly through changes in water resorption in the nephrons and collecting ducts, and partly by changes in glomerular filtration rate involving changes in the proportion of reptilian-type nephrons that are filtering.

The excretion of nitrogen as undissolved uric acid and urate salts frees osmotic space for excretion of salts and other solutes in urine. Substantial amounts of excreted cations may be combined with the undissolved urates. Salts such as NaCl and KCl account for only a moderate part of the osmotic concentration of the liquid fraction of the urine, and research is needed to characterize the principal osmolytes of avian urine.

The cloaca and rectum may play an important role in salt balance through resorption of NaCl from the urine of seed eaters that get little Na from their food. Potassium may be secreted into the urine by cloaca and rectum.

The nasal salt glands of birds having high salt intakes, such as sea birds, excrete concentrated NaCl solution, making water available to the organism. Nasal salt secretion is under control of parasympathetic innervation, and appears to be triggered by excitation of blood tonicity receptors.

REFERENCES

Akester, A. R., Anderson, R. S., Hill, K. J. and Osbaldiston, G. W. (1967). Radiographic study of urine flow in the domestic fowl. *Br. Poultry Sci.* **8**, 209–212.

Ames, E., Steven, K. and Skadhauge, E. (1971). Effects of arginine vasotocin on renal excretion of Na^+, K^+, Cl^- and urea in the hydrated chicken. *Am. J. Physiol* **221**, 1223–1228.

Bartholomew, G. A. and Cade, T. J. (1963). The water economy of land birds. *Auk* **80**, 504–539.

Bartholomew, G. A. and Dawson, W. R. (1953). Respiratory water loss in some birds of southwestern United States. *Physiol. Zool.* **26**, 162–166.

Bartholomew, G. A. and Trost, C. H. (1970). Temperature regulation in the speckled mousebird, *Colius striatus*. *Condor* **72**, 141–146.

Bartholomew, G. A., Hudson, J. W. and Howell, T. R. (1962). Body temperature, oxygen consumption, evaporative water loss, and heart rate in the poor-will. *Condor* **64**, 117–125.

Benedict, F. G. and Lee, R. C. (1937). "Lipogenesis in the Animal Body, with Special Reference to the Physiology of the Goose." Carnegie Institute of Washington, Publication Number 489, Washington, DC.

Bernstein, M. H. (1971). Cutaneous water loss in small birds. *Condor* **73**, 468–469.

Bindslev, N. and Skadhauge, E. (1971a). Salt and water permeability of the epithelium of the coprodeum and large intestine in the normal and dehydrated fowl (*Gallus domesticus*). In vivo perfusion studies. *J. Physiol. (Lond.)* **216**, 735–751.

Bindslev, N. and Skadhauge, E. (1971b). Sodium chloride absorption and solute-linked water flow across the epithelium of the coprodeum and large intestine in the normal and dehydrated fowl (*Gallus domesticus*). In vivo perfusion studies. *J. Physiol. (Lond.)* **216**, 753–768.

Braun, E. J. and Dantzler, W. H. (1972). Function of mammalian-type and reptilian-type nephrons in kidney of desert quail. *Am. J. Physiol.* **222**, 617–629.

Braun, E. J. and Dantzler, W. H. (1974). Effects of ADH on single-nephron glomerular filtration rates in the avian kidney. *Am. J. Physiol.* **226**, 1–8.

Burford, H. J. and Bond, R. F. (1968). Avian cardiovascular parameters: effect of intra-venous osmotic agents, relation to salt gland secretion. *Experientia* **24**, 1068–1088.

Cade, T. J. (1964). Water and salt balance in granivorous birds. "Thirst, Proceedings of the 1st International Symposium on Thirst in the Regulation of Body Water", pp. 237–256.

Cade, T. J. and Dybas, J. A. (1962). Water economy of the budgerygah. *Auk* **79**, 345–364.

Cade, T. J. and Greenwald, L. (1966). Nasal salt secretion in falconiform birds. *Condor* **68**, 338–350.

Cade, T. J., Tobin, C. A. and Gold, A. (1965). Water economy and metabolism of two estrildine finches. *Physiol. Zool.* **38**, 9–33.

Calder, W. A. (1964). Gaseous metabolism and water relations of the zebra finch, *Taeniopygia castanotis*. *Physiol. Zool.* **37**, 400–413.

Calder, W. A. and Bentley, P. J. (1967). Urine concentrations of two carnivorous birds, the white pelican and the roadrunner. *Comp. Biochem. Physiol.* **22**, 607–609.

Calder, W. A. and Schmidt-Nielsen, K. (1967). Temperature regulation and evaporation in the pigeon and the roadrunner. *Am. J. Physiol.* **213**, 883–889.

Carey, C. and Morton, M. L. (1971). A comparison of salt and water regulation in California quail (*Lophortux californicus*) and Gambel's quail (*Lophortyx gambelii*). *Comp. Biochem. Physiol.* **39A**, 75–101.

Chapman, T. E. and Black, A. L. (1967). Water turnover in chickens. *Poultry Sci.* **46**, 761–765.

Chapman, T. E. and McFarland, L. Z. (1971). Water turnover in coturnix quail with individual observations on a burrowing owl, Petz conure and vulturine fish eagle. *Comp. Biochem. Physiol.* **39A**, 653–656.

Chew, R. M. (1961). Water metabolism of desert-inhabiting vertebrates. *Biol. Rev.* **36**, 1–31.

Cooch, F. G. (1964). A preliminary study of the survival value of a functional salt gland in prairie anatidae. *Auk* **81**, 380–393.

Coulombe, H. N. (1970). Physiological and physical aspects of temperature regulation in the burrowing owl *Speotyto cunicularia*. *Comp. Biochem. Physiol.* **35**, 307–337.

Crawford, E. C., Jr. and Lasiewski, R. C. (1968). Oxygen consumption and respiratory evaporation of the emu and rhea. *Condor* **70**, 333–339.

Dantzler, W. H. (1966). Renal response of chickens to infusion of hyperosmotic sodium chloride solution. *Am. J. Physiol.* **210**, 640–646.

Dantzler, W. H. (1970). Kidney function in desert vertebrates. In "Hormones and the Environment" (G. K. Benson and J. G. Phillips, eds), pp. 157–190. Memoirs of the Society of Endocrinology Number 18. Cambridge University Press.

Dantzler, W. H. (1973). Characteristics of urate transport by isolated perfused snake proximal renal tubules. *Am. J. Physiol.* 224, 445–453.

Dawson, W. R. (1965). Evaporative water losses of some Australian parrots. *Auk* 82, 106–108.

Dawson, W. R. and Bartholomew, G. A. (1968). Temperature regulation and water economy of desert birds. In "Desert Biology" (G. W. Brown Jr., ed.). Academic Press, New York and London, pp. 357–394.

Diamond, J. and Bossert, W. H. (1968). Functional consequences of ultrastructural geometry in "backwards" fluid-transporting epithelia. *J. Cell. Biol.* 37, 694–702.

Dicker, S. E. and Haslam, J. (1972). Effect of exteriorization of the ureters on the water metabolism of the domestic fowl. *J. Physiol. (Lond.)* 224, 515–520.

Dixon, J. M. (1958). Investigation of urinary water reabsorption in the cloaca and rectum of the hen. *Poultry Sci.* 37, 410–414.

Ellis, R. A. (1965). DNA labelling and X-irradiation studies of the phosphatase-positive peripheral cells in the nasal (salt) glands of ducks. *Am. Zool.* 5, 648.

Emery, N., Poulson, T. L. and Kinter, W. B. (1972). Production of concentrated urine by avian kidneys. *Am. J. Physiol.* 223, 180–187.

Ernst, S. A. and Ellis, R. A. (1969). The development of surface specialization in the secretory epithelium of the avian salt gland in response to osmotic stress. *J. Cell. Biol.* 40, 305–321.

Fänge, R., Schmidt-Nielsen, K. and Osaki, H. (1958a). The salt gland of the herring gull. *Biol. Bull., Woods Hole* 115, 162–171.

Fänge, R., Schmidt-Nielsen, K. and Robinson, M. (1958b). Control of secretion from the avian salt gland. *Am. J. Physiol.* 195, 321–326.

Fletcher, G. L. and Holmes, W. N. (1968). Observations on the intake of water and electrolytes by the duck (*Anas platyrhynchos*) maintained on fresh water and on hypertonic saline. *J. exp. Biol.* 49, 325–339.

Folk, R. L. (1969). Spherical urine in birds: petrography. *Science N.Y.* 166, 1516–1519.

Gordon, M. S. (1972). "Animal Physiology: Principles and Adaptations" second edn. Macmillan, New York.

Greenwald, L., Stone, W. B. and Cade, T. J. (1967). Physiological adjustments of the budgerygah (*Melopsittacus undulatus*) to dehydrating conditions. *Comp. Biochem. Physiol.* 22, 91–100.

Grinnell, J. and Miller, A. H. (1944). "The Distribution of the Birds of California" Pacific Coast Avifauna Number 27, Cooper Ornithological Club, Berkeley.

Hanwell, A. and Peaker, M. (1973). The effect of post-ganglionic denervation on functional hypertrophy in the salt gland of the goose during adaptation to salt water. *J. Physiol. (Lond.)* 234, 78–80P.

Hanwell, A. and Peaker, M. (1975). The control of adaptive hypertrophy in the salt glands of geese and ducks. *J. Physiol. (Lond.).* 248 193–206.

Hanwell, A., Linzell, J. L. and Peaker, M. (1971a). Salt-gland secretion and blood flow in the goose. *J. Physiol. (Lond.)* 213, 373–387.

Hanwell, A., Linzell, J. L. and Peaker, M. (1971b). Cardiovascular response to salt-loading in conscious domestic geese. *J. Physiol. (Lond.)* 213, 389–398.

Hanwell, A., Linzell, J. L. and Peaker, M. (1972). Nature and location of the receptors for salt-gland secretion in the goose. *J. Physiol. (Lond.)* 226, 453–472.

Hart, W. M. and Essex, H. E. (1942). Water metabolism of the chicken with special reference to the role of the cloaca. *Am. J. Physiol.* 136, 657–668.

Hester, H. R., Essex, H. E. and Mann, F. C. (1940). Secretion of urine in the chicken, *Gallus domesticus*. *Am. J. Physiol.* **128**, 592–602.

Holmes, W. N. (1965). Some aspects of osmoregulation in reptiles and birds. *Archives Anat. Micr.* **54**, 491–514.

Holmes, W. N. (1972). Regulation of electrolyte balance in marine birds with special reference to the role of the pituitary-adrenal axis in the duck (*Anas platyrhynchos*). *Fed. Proc.* **31**, 1587–1598.

Holmes, W. N. and Stewart, D. J. (1968). Changes in nucleic acids and protein composition of the nasal glands from the duck (*Anas platyrhynchos*) during the period of adaptation to hypertonic saline. *J. exp. Biol.* **48**, 509–519.

Holmes, W. N. and Wright, A. (1969). Some aspects of the control of osmoregulation and homeostasis in birds. In "Progress in Endocrinology" (C. Gual and F. J. G. Ebling eds), Proc. IIIrd International Congress of Endocrinology, 1968, pp. 237–248.

Holmes, W. N., Fletcher, G. L. and Stewart, D. J. (1968). The patterns of renal electrolyte excretion in the duck (*Anas platyrhynchos*) maintained on fresh water and on hypertonic saline. *J. exp. Biol.* **48**, 487–508.

Hughes, M. R. (1972). Hypertonic salt gland secretion in the glaucous-winged gull, *Larus glaucescens*, in response to stomach loading with dilute sodium chloride. *Comp. Biochem. Physiol.* **41A**, 121–127.

Hughes, M. R. and Blackman, J. G. (1973). Cation content of salt gland secretion and tears in the brolga, *Grus rubicundus* (Perry) (Aves: Gruidae). *Aust. J. Zool.* **21**, 515–518.

Johnson, I. M. (1969). Electrolyte and water balance of the red-tailed hawk, *Buteo jamaicensis*. *Am. Zool.* **9**, 587.

Johnson, O. W. (1968). Some morphological features of avian kidneys. *Auk*. **85**, 216–228.

Johnson, O. W. (1974). Relative thickness of the renal medulla in birds. *J. Morph.* **142**, 277–284.

Johnson, O. W. and Mugaas, J. N. (1970a). Some histological features of avian kidneys. *Am. J. Anat.* **127**, 423–436.

Johnson, O. W. and Mugaas, J. N. (1970b). Quantitative and organizational features of the avian renal medulla. *Condor* **72**, 288–292.

Johnson, O. W. and Ohmart, R. D. (1973). Some features of water economy and kidney microstructure in the large-billed savannah sparrow (*Passerculus sandwichensis rostratus*). *Physiol. Zool.* **46**, 276–284.

Johnson, O. W. and Skadhauge, E. (1975). Structural-functional correlations in the kidneys and observations of colon and cloacal morphology in certain Australian birds. *J. Anat.* **120**, 495–505.

Johnson, O. W., Phipps, G. L. and Mugaas, J. N. (1972). Injection studies of cortical and medullary organization in the avian kidney. *J. Morph.* **136**, 181–190.

Johnson, R. E. (1968). Temperature regulation in the white-tailed ptarmigan, *Lagopus leucurus*. *Comp. Biochem. Physiol.* **24**, 1003–1014.

Kendeigh, S. C. (1939). The relation of metabolism to the development of temperature regulation in birds. *J. exp. Zool.* **82**, 419–438.

Kendeigh, S. C. (1944). Effect of air temperature on the rate of energy metabolism in the English sparrow. *J. exp. Zool.* **96**, 1–16.

Komnick, H. and Kniprath, E. (1970). Morphometrische Untersuchungen an der Salzdrüse von Silbermöwen. *Cytobiologie* **1**, 228–247.

Korr, I. M. (1939). The osmotic function of the chicken kidney. *J. cell. comp. Physiol.* **13**, 175–193.

Krag, B. and Skadhauge, E. (1972). Renal salt and water excretion in the budgerygah (*Melopsittacus undulatus*). *Comp. Biochem. Physiol.* **41A**, 667–683.

Lasiewski, R. C. (1963). Oxygen consumption of torpid, resting, active and flying hummingbirds. *Physiol. Zool.* **36**, 122–140.

Lasiewski, R. C. (1964). Body temperature, heart and breathing rate, and evaporative water loss in hummingbirds. *Physiol. Zool.* **37**, 212–223.

Lasiewski, R. C., Bernstein, M. H. and Ohmart, R. D. (1971). Cutaneous water loss in the roadrunner and poor-will. *Condor* **73**, 470–472.

Lee, P. and Schmidt-Nielsen, K. (1971). Respiratory and cutaneous evaporation in the zebra finch: effect on water balance. *Am. J. Physiol.* **220**, 1598–1605.

Lonsdale, K. and Sutor, D. J. (1971). Uric acid dihydrate in bird urine. *Science, N.Y.* **172**, 958–959.

Louw, G. N., Belonje, P. C. and Coetzee, H. J. (1969). Renal function, respiration, heart rate and thermoregulation in the ostrich (*Struthio camelus*). *Scient. Pap. Namib Des. Res. Stn. No. 42.* "Dr. Fitzsimons Commemorative Volume" (C. Koch, ed.), Namib Desert Research Association, Pretoria, pp. 43–54.

Lustick, S. (1970). Energetics and water regulation in the cowbird (*Molothrus ater obscurus*). *Physiol. Zool.* **43**, 270–287.

MacMillen, R. E. and Trost, C. H. (1967). Thermoregulation and water loss in the Inca dove. *Comp. Biochem. Physiol.* **20**, 263–273.

McNabb, F. M. A. (1969a). A comparative study of water balance in three species of quail—I. Water turnover in the absence of temperature stress. *Comp. Biochem. Physiol.* **28**, 1045–1058.

McNabb, F. M. A. (1969b). A comparative study of water balance in three species of quail—II. Utilization of saline drinking solutions. *Comp. Biochem. Physiol.* **28**, 1059–1074.

McNabb, F. M. A. and Poulson, T. L. (1970). Uric acid excretion in pigeons, *Columba livia. Comp. Biochem. Physiol.* **33**, 933–939.

McNabb, F. M. A., McNabb, R. A. and Steeves, H. R. III. (1973). Renal mucoid materials in pigeons fed high and low protein diets. *Auk* **90**, 14–18.

McNabb, R. A. (1974). Urate and cation interactions in the liquid and precipitated fractions of avaian urine, and speculations on their physico-chemical state. *Comp. Biochem. Physiol.* **48A**, 45–54.

McNabb, R. A. and McNabb, F. M. A. (1977). Avian urinary precipitates: their physical analysis, and their differential inclusion of cations (Ca, Mg) and anion (Cl). *Comp. Biochem. Physiol.* **56A**, 621–625.

McNabb, R. A., McNabb, F. M. A. and Hinton, A. P. (1973). The excretion of urate and cationic electrolytes by the kidney of the male domestic fowl (*Gallus domesticus*). *J. comp. Physiol.* **82**, 47–57.

Minnich, J. E. (1972). Excretion of urate salts by reptiles. *Comp. Biochem. Physiol.* **41A**, 535–549.

Minnich, J. E. and Piehl, P. A. (1972). Spherical precipitates in the urine of reptiles. *Comp. Biochem. Physiol.* **41A**, 551–554.

Moldenhauer, R. R. and Wiens, J. A. (1970). The water economy of the sage sparrow, *Amphispiza belli nevadensis. Condor* **72**, 265–275.

Mugaas, J. N. and Templeton, J. R. (1970). Thermoregulation in the red-breasted nuthatch (*Sitta canadensis*). *Condor* **72**, 125–132.

Nechay, B. R., Boyarsky, S. and Catacutan-Labay, P. (1968a). Rapid migration of urine into intestine of chickens. *Comp. Biochem. Physiol.* **26**, 369–370.

Nechay, B. R. and Carmen Lutherer, B. Del. (1968b). Handling of urine by cloaca and ureter in chickens. *Comp. Biochem. Physiol.* **26**, 1099–1105.

Ohmart, R. D. (1972). Physiological and ecological observations concerning the salt-secreting nasal glands of the roadrunner. *Comp. Biochem. Physiol.* **43A**, 311–316.

Ohmart, R. D., Chapman, T. E. and McFarland, L. Z. (1970a). Water turnover in roadrunners under different environmental conditions. *Auk* **87**, 787–793.

Ohmart, R. D., McFarland, L. Z. and Morgan, J. P. (1970b). Urographic evidence that urine enters the rectum and ceca of the roadrunner (*Geococcyx californianus*) Aves. *Comp. Biochem. Physiol.* **35**, 487–489.

Peaker, M. (1971). Avian salt glands. *Phil. Trans. Roy. Soc. Ser. B.* **262**, 289–300.

Peaker, M. and Linzell, J. L. (1975). "Salt Glands in Birds and Reptiles." Cambridge University Press, Cambridge.

Peaker, M., Peaker, S. J., Hanwell, A. and Linzell, J. L. (1973). Sensitivity of the receptors for salt-gland secretion in the domestic duck and goose. *Comp. Biochem. Physiol.* **44A**, 41–46.

Phillips, J. G. and Ensor, D. M. (1972). The significance of environmental factors in the hormone mediated changes of nasal (salt) gland activity in birds. *Gen. comp. Endocr. Suppl.* **3**, 393–404.

Poulson, T. L. (1965). Countercurrent multipliers in avian kidneys. *Science N.Y.* **148**, 389–391.

Poulson, T. L. and Bartholomew, G. A. (1962a). Salt balance in the savannah sparrow. *Physiol. Zool.* **35**, 109–119.

Poulson, T. L. and Bartholomew, G. A. (1962b). Salt utilization in the house finch. *Condor* **64**, 245–252.

Randle, H. W. and Dantzler, W. H. (1973). Effects of K^+ and Na^+ on urate transport by isolated perfused snake renal tubules. *Am. J. Physiol.* **225**, 1206–1214.

Schmidt-Nielsen, K. (1960). Salt-secreting nasal gland of marine birds. *Circulation* **21**, 955–967.

Schmidt-Nielsen, K., Jorgensen, C. B. and Osaki, H. (1957). Secretion of hypertonic solutions in marine birds. *Fed. Proc.* **16**, 113–114.

Schmidt-Nielsen, K., Jorgensen, C. B. and Osaki, H. (1958). Extrarenal salt excretion in birds. *Am. J. Physiol.* **193**, 101–107.

Schmidt-Nielsen, K., Borut, A., Lee, P. and Crawford, E. (1963). Nasal salt excretion and the possible function of the cloaca in water conservation. *Science N.Y.* **142**, 1300–1301.

Schmidt-Nielsen, K., Kanwisher, J., Lasiewski, R. C., Cohn, J. E. and Bretz, W. L. (1969). Temperature regulation and respiration in the ostrich. *Condor* **71**, 341–352.

Shoemaker, V. H. (1972). Osmoregulation and excretion in birds. "Avian Biology" (D. S. Farner and J. R. King, eds) Vol. II. Academic Press, New York and London, pp. 527–574.

Skadhauge, E. (1964). Effects of unilateral infusion of arginine-vasotocin into the portal circulation of the avian kidney. *Acta endocrin.* **47**, 321–330.

Skadhauge, E. (1967). *In vivo* perfusion studies of the cloacal water and electrolyte resorption in the fowl (*Gallus domesticus*). *Comp. Biochem. Physiol.* **23**, 483–501.

Skadhauge, E. (1968). The cloacal storage of urine in the rooster. *Comp. Biochem. Physiol.* **24**, 7–18.

Skadhauge, E. (1973). Renal and cloacal salt and water transport in the fowl (*Gallus domesticus*). *Danish Medical Bull.* **20**, Suppl. 1, 1–82.

Skadhauge, E. (1974a). Cloacal resorption of salt and water in the galah (*Cacatua roseicapilla*). *J. Physiol. (Lond.)* **240**, 763–773.

Skadhauge, E. (1974b). Renal concentrating ability in selected west Australian birds. *J. exp. Biol.* **61**, 269–276.

Skadhauge, E. (1975). Renal and cloacal transport of salt and water. "Advances in Avian Physiology" (M. Peaker, ed.) Symp. zool. Soc. Lond. No. 35, pp. 97–106.
Skadhauge, E. and Bradshaw, S. D. (1974). Saline drinking and cloacal excretion of salt and water in the zebra finch. *Am. J. Physiol.* **227**, 1263–1267.
Skadhauge, E. and Kristensen, K. (1972). An analogue computer simulation of cloacal resorption of salt and water from ureteral urine in birds. *J. theor. Biol.* **35**, 473–487.
Skadhauge, E. and Schmidt-Nielsen, B. (1967a). Renal medullary electrolyte and urea gradient in chickens and turkeys. *Am. J. Physiol.* **212**, 1313–1318.
Skadhauge, E. and Schmidt-Nielsen, B. (1967b). Renal function in domestic fowl. *Am. J. Physiol.* **212**, 793–798.
Smyth, M. and Bartholomew, G. A. (1966). Effects of water deprivation and sodium chloride on the blood and urine of the mourning dove. *Auk* **83**, 597–602.
Sperber, I. (1960). Excretion. "Biology and Comparative Physiology of Birds" (A. J. Marshall, ed.). Academic Press, New York and London, pp. 469–492.
Stephenson, A. H. (1974). Seasonal variations in water and energy turnover in song sparrows in southeastern Wisconsin. M.Sc. Thesis, University of Wisconsin, Milwaukee.
Stewart, D. J. (1972). Secretion by salt gland during water deprivation in the duck. *Am. J. Physiol.* **223**, 384–386.
Sturkie, P. D. (1965). "Avian Physiology" second edn. Comstock Publishing Association, Cornell University Press, Ithaca, pp. 372–405.
Trost, C. H. (1968). Adaptations of horned larks to stressful environments. Ph.D. Thesis, University of California, Los Angeles.
Utter, J. M. and LeFebvre, E. A. (1973). Daily energy expenditure of purple martins (*Progne subis*) during the breeding season: estimates using D_2O^{18} and time budget methods. *Ecology* **54**, 595–604.
Verney, E. B. (1947). The antidiuretic hormone and factors which determine its release. *Proc. roy. Soc. Lond. Ser. B*, **135**, 27–106.
Willoughby, E. J. (1968). Water economy of the Stark's lark and grey-backed finch-lark from the Namib Desert of South West Africa. *Comp. Biochem. Physiol.* **27**, 723–745.
Willoughby, E. J. (1969). Evaporative water loss of a small xerophilous finch, *Lonchura malabarica*. *Comp. Biochem. Physiol.* **28**, 655–664.
Willoughby, E. J. (1971). Metabolism of inorganic cations by quail (*Coturnix coturnix*) drinking solutions of $CaCl_2$ and $MgCl_2$. *Comp. Biochem. Physiol.* **38A**, 541–554.
Zaks, N. G. and Sokolova, M. M. (1961). Ontogenetic and specific features of the nasal gland in some marine birds. *Sechenov Physiol. J. U.S.S.R.* **47**, 120–127.

2. Australian Marsupials and Monotremes*

P. J. BENTLEY

University of New York, New York, USA

I.	Introduction	57
	A. The Geography, Vegetation and Climate of Australia	59
II.	Marsupials	61
	A. Problems of Osmoregulation	61
	B. Pulmocutaneous Evaporation in Marsupials	63
	C. The Kidneys	67
	D. Regulation of Renal Function	72
	E. The Role of the Gut in Osmoregulation	74
	F. Effects of Reproduction on Water and Salt Metabolism	75
	G. Behaviour and Osmoregulation	77
	H. Effects of the Availability of Water on the Life of Marsupials	81
III.	Monotremes	85
	A. Distribution	85
	B. Evaporative Water Losses	86
	C. Kidney Function	87
	D. Effects of Environment and Behaviour on Osmoregulation	88
IV.	Conclusions	89
	References	90

I. INTRODUCTION

Most species of the extant marsupials and all of the monotremes are confined to the Australasian region where they are the most interesting and predominant mammals. Some placentals, notably rodents, bats and a wild dog (the dingo) also were living there before the arrival of Europeans. These mammals, however, do not appear to have adversely

---

* This chapter is dedicated to Professor Harry Waring who first suggested that marsupial osmoregulation was a proper subject for young people to investigate.

influenced the adaptation and survival of the marsupials that preceded them there. The marsupials and monotremes have considerable systematic interest as they phylogenetically diverged from the placental mammals in the Cretaceous period about 100 million years ago. Some marsupials have persisted in the Americas but about 75 of the 90 or so extant genera are found in Australia and New Guinea. All the extant, and extinct, monotremes are natives of Australasia. The only unique physiological difference among these three subclasses of the Mammalia is in their methods of reproduction. The monotremes are oviparous, while only the relatively early stages of fetal development take place *in utero* in marsupials which lack a "true" or allantoic placenta. Compared to placentals (or Eutherians), marsupials are born in a relatively immature condition and subsequent development takes place in an external pouch or marsupium.

Different species of monotremes and marsupials exhibit many other interesting physiological characters but when these are compared to the placentals none of them are really unique. Possibly the most widespread physiological character, apart from reproductive processes, which is not shared with the majority of placentals, is a somewhat low body temperature (Sutherland, 1897; Martin, 1903; Robinson, 1954). This character is accompanied by a basal metabolic rate which is about 30% lower than in placentals (Martin, 1903; Schmidt-Nielsen, Dawson and Crawford, 1966; Dawson and Hulbert, 1969; MacMillan and Nelson, 1969; Arnold and Shield, 1970). These two interesting physiological characters, which are probably related to each other, may have important effects on the animal's requirements for water.

Apart, however, from understanding the effects of such physiological differences we may ask why the study of marsupial osmoregulation is of any particular interest. There are several answers which are of both academic and practical importance. The long geographic and genetic isolation of the monotremes and marsupials from the placentals provides the prospect of an evolution of different osmoregulatory mechanisms. The Australian marsupials have in their island isolation, evolved numerous species, ranging in weight from a few grams to 100 kg, which are adapted to life in a vast number of habitats, and which show many general morphological similarities to placentals that live in comparable situations elsewhere. The common names of marsupials may illustrate these similarities. There are thus so-called marsupial mice, marsupial cats, hare-wallabies, a marsupial mole, Tasmanian wolf and even the Tasmanian devil. Marsupials may be herbivorous, carnivorous or omnivorous in their feeding habits. Some live by predation, others graze in open grasslands or woodlands, while many are arboreal and even fossorial in their manner of life. Australia is a very large island which provides a variety of habitats that range from hot, dry deserts to tropical rain forest and even regions of winter snow. The local patterns of the seasons also vary and include periodic rains and changes in the environmental temperature which are often quite unpredictable. The Australian marsupials and monotremes in their antipodean isolation thus provide a naturally sequestered laboratory for studying the diverse interactions and adaptations that have taken place in a single closely related mammalian order. Of more practical interest is the conservation of these interesting and unique mammals and their commercial utilization. The latter is especially important with respect to the herbivorous species which may compete for food with sheep and cattle and have thus often been given the ignominious title of pests. The justification, necessity and practical measures for such pest control are often intimately related to the occurrence of periodic droughts and the animal's needs for water.

A. THE GEOGRAPHY, VEGETATION AND CLIMATE OF AUSTRALIA

It was mentioned earlier that Australia offers a wide range of habitats in which animals can live and these place different stresses on their capacities to physiologically regulate their water and salt balance. Australia stretches from a northern tropical equatorial zone starting at about 11°S and extending to 40°S, an area where snow may fall in the wintertime. From west to east, longitudes 113° to 154°, the island is about 4000 km across. The northern regions of this continent are tropical, the south is temperate while the interior two-thirds is mostly hot and dry with a periodic but unpredictable average annual rainfall of less than 25 cm (see Fig. 1). The central area is largely open plain broken in some parts by low hills and outcrops of rock. Much of it is covered, though usually rather sparsely, with grasses and low scrub in which natural water supplies are limited and infrequent. In this interior region the air temperature may regularly exceed 40°C and as the accompanying relative humidity is usually low, the potential evapotranspiration is high. There nevertheless remain large areas of the continent where water is relatively plentiful, such as in the tropical north and in the more agricultural areas along the east coast and in the south-east and south-west. All these regions are occupied by marsupials though there are differences in their population densities and in the species distribution there. This distribution is illustrated in Fig. 1, which shows the areas occupied by the Macropodidae.

Apart from water, there is also considerable variation in the salt content of the vegetation. Animals living in the mountainous plateaux of the south-east exhibit a salt appetite due to the low salt content of the plants that they eat. Large areas of salt lakes, which are usually dry, also exist in many parts of Australia and these may support halophytic vegetation which can provide food but because of its high salt content could have osmotic consequences on the animals eating it.

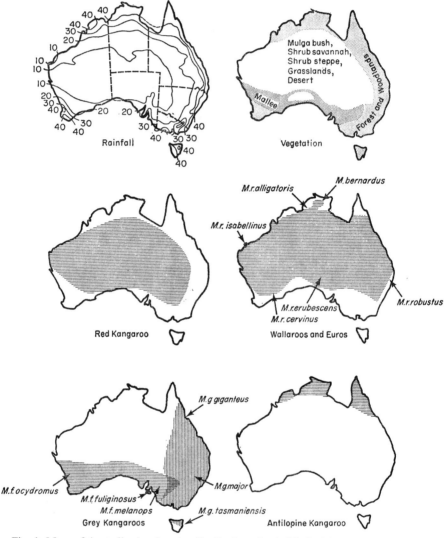

Fig. 1. Maps of Australia showing the distribution of rainfall (in in), the general type of vegetation and species and sub-species of the Macropodidae. Red kangaroos, *Megaleia rufa*; Wallaroos and Euros, *Macropus robustus*, antilopine kangaroos, *Macropus antelopinus* and grey kangaroos, *Macropus giganteus* is the eastern grey kangaroo (vertical hatching) and *Macropus fuliginosus* is the western grey kangaroo (horizontal hatching) (from Frith and Calaby, 1969).

II. MARSUPIALS

A. PROBLEMS OF OSMOREGULATION

The avenues for the gain, loss and conservation of water and salts are basically the same for marsupials (and monotremes) as those of other mammals. Water is principally lost by evaporation from the skin and pulmonary tract but also in the urine and faeces. Individual quantitative variations exist between species but generally, possibly with a single exception related to their lower metabolic rate, the differences among marsupials encompass the same range that is seen among placental mammals. However, less information is available about marsupial osmoregulation.

The water turnover of several species of marsupials has been estimated by measuring changes in the body's levels of tritiated water (Richmond et al., 1962). These results have been summarized in Table I. Large animals, with their relatively lower rates of metabolism and surface area, will be expected to have a lower water turnover than small animals. It is considered that if the turnover is corrected to the 0·8 power of the body weight that most of the effect of the animal's size will be compensated for, so any remaining difference will then reflect other physiological factors and adjustments (Table I). It is not yet clear whether or not the lower rate of metabolism in marsupials reduces their water turnover as compared to placentals. Among the Macropodidae (kangaroos), however, it does appear to have this effect. Nevertheless a comparison of the water turnover of Australia dasyurids (marsupial mice) with that of native rodents (Muridae) failed to show any clear differences (Haines et al., 1974). Indeed, although the rates of VO_2 consumption were, as expected, lower in the marsupials, the rates of water loss were similar in the two groups so that the dasyurids expended more water ml^{-1} O_2 consumed than the placentals.

When the environmental temperature is raised or if marsupials are lactating, the water turnover is increased but if drinking water is restricted it declines. The latter reduction reflects the utilization of regulatory mechanisms that conserve water.

Marsupials that live in hot, arid areas will be expected to experience potentially greater ionic and osmotic problems than their relatives from more temperate regions. This difference principally reflects additional evaporative water loss including that used for temperature regulation. In addition, however, supplies of drinking water are also usually more limited in such hot, dry climates and the vegetation that is eaten by the herbivorous species may contain less water. Seasonal

TABLE I

Water turnover (measured with tritiated water) in marsupials

Species	Wt in g	ml kg^{-1} day^{-1}	(ml kg$^{0.8}$)$^{-1}$ day^{-1}	% body H$_2$O day	Reference
Macropodidae					
Megaleia rufa (red kangaroo)	(40 kg^4)	88	—	—	Macfarlane et al. (1963)
Petrogale inornata (rock wallaby)	3180	81	99	11·8	Kennedy and Heinsohn (1974)
Phalangeridae					
Trichosurus vulpecula (brush possum)	1520	89	96	13·0	Kennedy and Heinsohn (1974)
Dasyuridae					
Sminthopsis crassicaudata (fat-tailed sminthopsis)	17	461	224	71·1	Macfarlane et al. (1971), Kennedy and Macfarlane (1971)
Dasycercus cristicauda (mulgara)	87	134	87	20·0	Macfarlane et al. (1971), Kennedy and Macfarlane (1971)
Dasyuroides byrnei (marsupial rat)	130	125	93	13·7	Macfarlane et al. (1971), Kennedy and Macfarlane (1971)
Peramelidae					
Isoodon macrourus (short-nosed bandicoot)	1468	132	98	23·3	Hulbert and Dawson (1974a)
Macrotis lagostis (rabbit-eared bandicoot)	1081	47	46	8·3	Hulbert and Dawson (1974a)
Perameles nasuta (long-nosed bandicoot)	972	68	73	10·9	Hulbert and Dawson (1974a)

The measurements were made using tritiated water under resting conditions with water available *ad libitum*. The environmental temperatures were 20 to 25°C.

droughts which are sometimes unpredictable and can last for several years also have a profound influence on the animal's survival and ability to reproduce. Even if such conditions do not immediately affect the supplies of drinking water, the food which is available will be reduced in amount and the animals may have to walk much farther to get it. Many marsupials from such arid areas are carnivores so that their osmotic problems are somewhat different as they may get adequate water from their food, though their survival will also be ultimately threatened by drought conditions.

In the succeeding sections the physiological and behavioural factors involved in the gains and losses of water and salts by marsupials (and monotremes) will be described. The osmoregulation of these animals in their natural environments is of special interest and this will be emphasized.

B. PULMOCUTANEOUS EVAPORATION IN MARSUPIALS

Evaporation from the marsupial integument, the skin and pulmonary tract, like that of placentals constitutes a major avenue for their losses of water. In dasyurid marsupial mice with water available *ad libitum*, it amounts to 30 to 40% of the total water loss each day (Haines *et al.*, 1974) and in a somewhat larger macropodid, the quokka, *Setonix brachyurus*, it is similar to this (Bentley, 1953, 1955). However, in three species of perameloid marsupials (bandicoots) evaporation was found to be much higher, contributing about 70% of the total water loss (Hulbert and Dawson, 1974a). This proportion may be expected to vary in a single species depending on the environmental conditions, such as air temperature and relative humidity, the animal's state of bodily hydration, its activity and metabolic rate.

Part of the evaporation from the integument is unavoidable, or obligatory, and reflects the saturation deficit for water vapour in the external air, the animals normal oxygen requirements and its body temperature. In addition, as marsupials are homeotherms, they may also utilize the evaporation of water to lower their body temperature if it should rise due to excessive heat gain from the environment or as a result of an accumulation of metabolic heat. Such increases in evaporative water loss in marsupials exposed to hot atmospheres have been shown to occur in a wide variety of marsupials (see for instance Robinson, 1954) though considerable interspecific differences in the responses exist.

There are very few observations in marsupials on the partition of evaporative water losses between the skin and the pulmonary tract. Bennetts wallaby, *Macropus rufogriseus*, at 25°C loses about 70% of its water by evaporation from the head region and this loss, presumably, is principally from the pulmonary tract (Gubbins and Johnson, 1972). In the red kangaroo, *Megaleia rufa*, at rest at 24°C, cutaneous and pulmonary evaporation contribute about equally to evaporation from the animal but during exercise, when sweating occurs, the contribution of the skin rises to about 60% (Dawson, 1973; Dawson et al., 1974).

When marsupials are exposed to thermal stress, so that they need to thermoregulate, their respiratory rate and concomitant evaporation increase. This response has been shown very clearly in an extensive survey of 25 species (Robinson and Morrison, 1957). It was noted that considerable interspecific differences exist; the responses being greatest in macropods and least in the small dasyurids. More direct measurements have since shown that considerable increases in pulmonary water loss occur in response to a rise in the environmental temperature in the kangaroos, *Megaleia rufa*, *Macropus robustus*, *Macropus eugenii* and the possum, *Trichosurus vulpecula* (Dawson, 1969; Dawson et al., 1969; Dawson, 1973). Three species of bandicoots have also been studied; those that inhabit temperate areas, *Isoodon macrourus* and *Perameles nasuta*, increased their respiratory rates and evaporation when exposed to hot atmospheres (Hulbert and Dawson, 1974b). However, the rabbit-eared bandicoot, *Macrotis lagotis*, which lives in dry areas, did not significantly change its rate of evaporation in response to such conditions (see also Robinson and Morrison, 1957). It has been suggested that this lack of a response may be an adaptation to life in dry areas where such an expenditure of water could be a disadvantage. In its natural desert habitat the rabbit-eared bandicoot avoids much of the heat of the day by burrowing into the ground.

Water loss from the skin may be percutaneous or be secreted by sweat glands. Primate-type eccrine sweat glands have been identified on the fore and hind paws of several species of marsupials (Bentley, 1955; Green, 1961). The rest of the body is, however, covered with apocrine sweat glands which are associated with the hair follicles. In many non-primates, these function as true sweat glands. Their role in marsupials has, however, not been extensively studied except in the macropodids, *Macropus rufogriseus*, and *M. rufa* (Gubbins and Johnson, 1972; Dawson et al., 1974). In these kangaroos such skin glands have been shown to mediate the increased cutaneous evaporation which is observed during exercise. These sweat glands appear to be under the control of adrenergic nerve fibres. They have not been shown to secrete in response to simple thermoregulatory need.

Evaporative water loss from the skin has been directly measured in several marsupials (Table II). An increased water loss occurs as the environmental temperature is raised but the change is not of a magnitude

TABLE II

Cutaneous water loss in marsupials (compared to sheep, with and without sweat glands)

Species	g H$_2$O m^{-2} h^{-1}				Reference
	20°C	25°C	40°C		
Macropodidae					
Macropus rufogriseus (Bennett's wallaby)	—	4.0	(20 at 45°C)		Gubbins and Johnson (1972)
Megaleia rufa (red kangaroo)	—	8.2	13.2		Dawson (1974)
Macropus robustus (euro)	—	8.6	15.0		Dawson (1974)
Dasyuridae					
Sarcophilus harrisii (Tasmania devil)	24.3	—	36.8		Hulbert and Rose (1972)
Peramelidae					
Isoodon macrourus (short-nosed bandicoot)	5.1	10.8	28.5		Hulbert and Dawson (1974a)
Macrotis lagotis (rabbit-eared bandicoot)	6.8	11.6	—		Hulbert and Dawson (1974a)
Peramelese nasuta (long-nosed bandicoot)	11.8	14.0	—		Hulbert and Dawson (1974a)
Sheep					
With no sweat glands	12.4	—	31.0		From Hulbert and Dawson (1974a)
Normal with sweat glands	22.6	—	63.1		From Hulbert and Dawson (1974a)

that is consistent with a thermoregulatory response. It is indeed similar to that observed in a strain of sheep that have an inherited absence of sweat glands. The precise pathway of this cutaneous water loss in marsupials is unknown but it could be the result of a minimal secretion from the apocrine sweat glands or be due to passive diffusion of water through the skin. The nature of the fur could also influence cutaneous evaporation due to its ability to limit direct heating of the skin surface and to trap water vapour in its interstices and so maintain an unstirred layer of water-saturated air. The latter could slow evaporation. It has been found that the fur of the red kangaroo, *M. rufa*, reflects more solar radiation than that of the euro, *M. robustus*, and its insulative properties are also greater (Dawson and Brown, 1970). However, as seen in Table II the cutaneous evaporation from both species is similar under laboratory conditions though it is unknown if this situation persists under more exposed conditions in the field.

Many marsupials have been observed to lick their fur, especially on their forelimbs, when they are exposed to thermal stress. This process is usually thought to involve the spreading of saliva on the fur but the magnitude of such water loss, and its importance in temperature regulation is uncertain. When fur licking was prevented, with the aid of a wide collar, in a small macropod, the quokka, *S. brachyurus*, no change in its ability to thermoregulate at 40°C was observed(Bentley, 1960). The importance of fur licking has also been discounted in *M. eugenii* and *T. vulpecula* (Dawson, 1969; Dawson et al., 1969). More recent studies, however, on *M. rufa* (Needham et al., 1974) suggest that the saliva could provide a significant amount of fluid for evaporative cooling from the fore limbs. The blood supply to this area can be greatly increased in response to exposure to hot atmospheres and this could assist heat loss. Licking does not occur in response to hot atmospheres in dehydrated *M. robustus* and *M. rufa*, yet they still thermoregulate, suggesting that it is not of crucial importance (Dawson, 1973). However, the bandicoot *I. macrourus* salivates profusely when exposed to a temperature of 40°C; this water constitutes about 40% of its total loss and appears to help the animal maintain a normal body temperature (Hulbert and Dawson, 1974b). The deposition of saliva on the fur may thus contribute to temperature regulation but its relative importance is uncertain and may differ between species and vary with the particular physiological and environmental conditions.

Evaporative water loss from marsupials may decrease under a variety of conditions including starvation, dehydration and torpor. As described above, normal evaporative water loss may be less in some marsupials than seen in comparable placentals and this reduction may reflect their lower metabolic rate. It has also been observed that, like certain placentals, some marsupials that live in deserts may have a lower rate of energy use which may help reduce the rate of water loss. This reduction most likely occurs in the evaporation from the pulmonary tract due to the decreased need for oxygen. A lower metabolic rate has been observed in the desert-dwelling mulgara (or marsupial mouse) *Dasycercus cristicauda* (MacMillen and Nelson, 1969; Kennedy and Macfarlane, 1971; Macfarlane *et al.*, 1971) and the rabbit-eared bandicoot *M. lagostis* (Hulbert and Dawson, 1974c). In the quokka, *S. brachyurus*, evaporative water loss is reduced following dehydration. This response, which is similar to that of the camel and laboratory rat (Bentley, 1960), has also been observed in Australian native rodents but it is absent in the marsupial-rat *Dasyuroides byrnei* (Haines *et al.*, 1974). Starvation may result in a reduction of the metabolic rate and the pulmonary evaporation of placentals but this possibility does not appear to have been investigated in marsupials. A shortage of food is, however, thought to be one of the factors initiating the hypothermia and torpor which is seen in several species of marsupials including the mulgara, *Dasycercus cristacauda*; the marsupial-mouse, *Sminthopsis crassicaudata*, and the pigmy possum, *Cercartetus nanus* (see, for instance, Bartholomew and Hudson, 1962; Kennedy and Macfarlane, 1971). Other factors, including behaviour, also influence the evaporation of water in marsupials and these will be discussed in subsequent sections.

C. THE KIDNEYS

As in all vertebrates, the kidneys of marsupials are of basic importance for their osmoregulation. Depending on the animal's diet and its environment, the kidney mediates the excretion or conservation of water, salts and urea. The kidneys in marsupials do not appear to be remarkable in any respect when compared to those of placentals (Bentley, 1955; Reid and McDonald, 1968). They exhibit a similar range of abilities to respond to different physiological and environmental situations.

Morphologically the kidneys of marsupials are similar to those of placentals; there is a glomerulus, a juxtaglomerular apparatus and a thin loop of Henle in the tubule. Like in placentals the latter is presumably associated with the animal's ability to form urine which is hyperosmotic to the plasma. There is also a zonation of the renal tissue into a cortex and medulla and the latter may be extended into a renal papilla. The relative thickness of the medulla varies interspecifically in a manner as shown in placentals, which indicates that the thicker it is, the greater the animal's ability to concentrate its urine (see, for instance, Schmidt-Nielsen and Newsome, 1962; Kinnear *et al.*, 1968;

Hulbert and Dawson, 1974a). The possession of a typical placental-type kidney by the marsupials is possibly somewhat surprising considering the long time that the two groups have been seperated from each other. This similarity could be the result of parallel evolution or, more likely, reflect the possession of such a structure by a common ancestor.

The only detailed study of marsupial glomerular and renal tubular function has been made on the brush possum *T. vulpecula* (Reid and McDonald, 1968a). Some of these results are summarized in Table III. It can be seen that the renal functions of the possum are comparable to those of similar-sized placentals such as rabbits and dogs.

The contribution of the urine in marsupials to their total water loss varies under different conditions just as is seen in placentals. Thus on a dry diet with water available *ad libitum*, (the environmental temperature was 21 to 25°C) the urine contributes 25 to 60% of the total water loss from the quokka *S. brachyurus* (P. J. Bentley, personal observations), 9 to 15% in bandicoots (Hulbert and Dawson, 1974a) and about 20% in the wallaby *M. eugenii* (Purohit, 1971). When the supply of drinking water is restricted, the flow of urine is reduced by as much as 80% and its concentration rises considerably. Urine concentrations show interspecific differences and may exceed 4000 mosmol kg H_2O^{-1} (Table IV) which is more than 10 times as concentrated as the plasma. This ability to concentrate the urine, which is comparable to that observed in placentals, results in a conservation of water, the importance of which varies with the environmental conditions that determine other losses of water, such as by evaporation and in the faeces. Such an ability to form a hyperosmotic urine also may make it possible for the animal to drink saline solutions and make a net gain of water which can be used for other bodily needs. Saline drinking will be discussed in Section II, G.3.

Several studies on field populations have shown that marsupials in the field fully use their renal abilities to limit the rate of secretion and to concentrate the urine. Thus the urine flows of *S. brachyurus* on their native Rottnest Island were about 25 ml h^{-1} during the cool wet winter but less than 3 ml h^{-1} in the hot, dry summer period (Bentley, 1955). Similarly, in the Kangaroo Island wallabies, *Macropus eugenii*, urine flows were found to be reduced by about 80% during the dry season, as compared to the wet season (Barker, 1971). Such decreases in urine flow are accompanied by increases in its concentrations which have also been shown to reflect seasonal conditions in the quokka, the euro, *M. robustus*, and the red kangaroo, *M. rufa* (Bentley, 1955; Ealey et al., 1965; Dawson and Denny, 1969).

The kidney is the major avenue for urea excretion and this may have

TABLE III

Renal clearances in a marsupial, *Trichosurus vulpecula* and a monotreme *Tachyglossus aculeatus* compared to two placental mammals of a similar size

Species	Clearances (ml min^{-1} kg^{-1})			Filtration fraction	Extraction PAH%
	Creatinine	Inulin	PAH		
Trichosurus vulpecula (brush possum)	3·32	2·52	12·99	0·2	89
Tachyglossus aculeatus (echidna)	1·17	—	—	—	—
Rabbit	3·12	—	18·2	0·17	—
Dog	4·29	—	13·51	0·29	90

The data for the possum are from Reid and McDonald (1968), the echidna from Bentley and Schmidt-Nielsen (1967) and the rabbit and dog from Smith (1951).

TABLE IV

Maximal recorded urinary concentrations in various species of marsupials and a monotreme

Species	Na + K mEq l^{-1}	Urea mmol l^{-1}	Osmolality mosmol kg H$_2$O^{-1}	Reference
Marsupials				
Megaleia rufa (red kangaroo)	≈750	≈250	2700	Dawson and Denny (1969), A. E. Newsome quoted by Schmidt-Nielsen (1964)
Macropus robustus (euro)	990	≈700	2730	Ealey *et al.* (1965)
Macropus eugenii (tammar)	1382 (Na)	—	3231	Purohit (1971)
Setonix brachyurus (quokka)	801	879	2200	Bentley (1955)
Trichosurus vulpecula (possum)	—	—	1504	Reid and McDonald (1968)
Dasycercus cristicauda (mulgara)	674	2610	4018	Schmidt-Nielsen and Newsome (1962)
Isoodon macrourus (bandicoot)	—	—	3126	Hulbert and Dawson (1974a)
Macrotis lagostis (bandicoot)	—	—	3976	Hulbert and Dawson (1974a)
Perameles nasuta (bandicoot)	—	—	4175	Hulbert and Dawson (1974a)
Monotreme				
Tachyglossus aculeatus (echidna)	≈400	—	2293	Bentley and Schmidt-Nielsen (1967)

important consequences on the animal's osmoregulation and nutrition. When the protein content of the diet is high, so that the basal bodily needs for it are exceeded, then larger amounts of urea appear in the urine. This excretion will result in an additional need for water. Such urinary urea is especially apparent in carnivorous species such as the mulgara, *D. cristacauda*, in which it may make up over 60% of the total solutes present in the urine (Table IV). In herbivorous species this proportion appears to depend on the nature of the diet (Barker, 1971; Ealey *et al.*, 1965; Dawson and Denny, 1969). When the protein content in the food is low, urea may contribute less than 10% of the total solutes in the urine of the kangaroos *M. robustus* and *M. rufa* (Ealey *et al.*, 1965; Dawson and Denny, 1969). However, this level may be much higher when the dietary grasses are richer in protein.

Urea may be recycled back into the gut of macropodid marsupials where it is used by the microflora to synthesize more protein (Brown, 1969). This process has also been observed in other herbivores such as cattle, camels and sheep. The kidney appears to play an important role in limiting the urea excretion in the urine of such species. In *M. eugenii* fed on a low-protein diet the urinary excretion of urea was small. The concentration of this solute in the renal tissue water declines from the kidney cortex to the medulla suggesting that a tubular conservation mechanism may be functioning (Lintern and Barker, 1969).

The kidney is also an organ for the conservation and excretion of salts, especially Na and K. These ions are normally obtained in the diet which may contain them in different amounts. The Na^+ concentration in the urine of the grey kangaroo, *Macropus giganteus*, and the wombat, *Vombatus hirsutus*, was found to be almost negligible in animals living in the Australian Alps (Blair-West *et al.*, 1968). In this mountainous area, the soil is poor in Na and this deficiency is reflected in the composition of the dietary plants. Such low urinary Na^+ concentrations contrast with those of animals living near the coast where the levels of Na in the soil are much higher. The kidney can clearly reduce the excretion of Na in the urine. In contrast to this situation some marsupials may gain an excess of salts in their diet as a result of the consumption of halophytic vegetation or by drinking brackish water. Marsupials have been shown to have an ability to excrete salt in their urine in concentrations that may even exceed that in sea water ($= 560$ mEq l^{-1} NaCl) (Table IV). It is possible that if they can excrete sufficient of the ingested salt, at these high concentrations, they could make a net gain of water from salty vegetation or brackish water. This ability seems to be quite well developed in the wallaby *M. eugenii* (Kinnear *et al.*, 1968; Purohit, 1971) and to a lesser extent in the quokka *S. brachyurus* (Bentley, 1955). These animals can even drink sea water but they do not appear to be able to gain enough water from this source to survive indefinitely with no fresh water to drink. Such drinking may, however, help to prolong their survival for short periods of time when supplies of fresh water are limited.

D. REGULATION OF RENAL FUNCTION

The excretion of water and salts by the kidneys are controlled in placentals by the antidiuretic hormone (ADH) that is secreted by the neurohypophysis, and corticosteroid hormones from the adrenal cortex. The latter gland is in turn controlled by the trophic effects of ACTH from the adenohypophysis and renin from the juxtaglomerular apparatus in the kidney. Renin appears to be more important for controlling the secretion of aldosterone which regulates Na and K excretion. In marsupials, the same basic processes appear to occur though there is at present less information available about this.

The amino acid composition of the neurohypophysial hormones show considerable variation among the vertebrates. However, the hormones present in marsupials appear to be identical to those in most placentals. 8-Arginine-vasopressin and oxytocin have been identified in *Trichosurus vulpecula*, *S. brachyurus* and *M. rufa* (Ferguson and Heller, 1965). The injection of commercial preparations of pitressin, which contains arginine-vasopressin, has an antidiuretic effect and increases the urine concentration in *S. brachyurus* (P. J. Bentley, personal observations; Bentley and Shield, 1962). An antidiuretic substance, which appears to be ADH has also been found in the blood of *Trichosurus caninus*, *Isoodon obesulus* and *S. brachyurus* when they are exposed to hot, dehydrating conditions (Robinson and Macfarlane, 1957). It thus seems likely that the ability to control urine flow in marsupials results from the action of an antidiuretic hormone which is probably the same as that in most placentals.

Adrenalectomy, in the quokka, *S. brachyurus*, results in death within a few days and is accompanied by a decline in the Na and a rise in the K concentrations in the plasma (Buttle *et al.*, 1952). The survival of these quokkas could be prolonged by the injection of deoxycorticosterone. A similar response to adrenalectomy has been observed in the brush possum, *T. vulpecula*, and the red kangaroo, *M. rufa*, in which the plasma electrolyte changes were also shown to be prevented by injected cortisol and aldosterone (Reid and McDonald, 1968; McDonald, 1974). This response is typical of that seen in placentals and reflects the ability of the adrenocortical hormones, especially aldosterone, to limit urinary Na loss, while promoting the excretion of K. The adrenocortical steroids cortisol, corticosterone and aldosterone have

TABLE V

Principal adrenocorticosteroids in marsupials and monotremes

Species	Cortisol (F)	Corticosterone (B)	Aldosterone	F/B ratio	Reference
Marsupials					
Megaleia rufa (red kangaroo)	+	+	+	50–250	Weiss and McDonald (1967)
Macropus giganteus (grey kangaroo)	+	+	+	5·5–28	Coghlan and Scoggins (1967)
Setonix brachyurus (quokka)	+	+	<0·1 µg ml⁻¹	1·1	Ilett (1969)
Trichosurus vulpecula (brush possum)	+	+	<0·1 µg ml⁻¹	3·5–12·8	Weiss and McDonald (1966b)
Dasyurus viverrinus (native cat)	+	+	+	5	Weiss and Richards (1971)
Sarcophilus harrisii (Tasmanian devil)	+	+	+	2·5	Weiss and Richards (1971)
Vombatus hirsutus (wombat)	+	+	+	1·8–12	Coghlan and Scoggins (1967); Weiss and McDonald (1966a)
Phascolarctos cinereus (koala)	+	+	+		Weiss and Richards (1970)
Perameles nasuta (long-nosed bandicoot)	+	?	?		Weiss and Richards (1971)
Monotremes					
Tachyglossus aculeatus (echidna)	+	+	+	0·1–0·4	Weiss and McDonald (1965); Weiss (1973)
Ornithorhynchus anatinus (platypus)	+	+	+	≈3·0	Weiss (1973)

been identified in the blood of a variety of Australian marsupials (Table V). A high ratio of cortisol to corticosterone is usually seen and appears to be characteristic of marsupials.

The levels of aldosterone in the blood of the grey kangaroo, *M. giganteus*, and the wombat, *V. hirsutus*, were found to be elevated in animals that were caught in the Na-deficient regions of south-east Australia (Coghlan and Scoggins, 1967; Blair-West et al., 1968). This increased activity of the adrenal cortex is associated with a hyperplasia of the zona glomerulosa, which is the aldosterone-producing tissue in the adrenal gland (Blair-West et al., 1968). It thus seems likely that aldosterone normally promotes the conservation of Na by increasing its reabsorption across the kidney tubules, just as in placentals.

Aldosterone secretion in placentals is stimulated as a result of the release of renin by the juxtaglomerular apparatus. This system exists in the marsupials and renin has been identified in the blood of eight species (Reid and McDonald, 1969; Simpson and Blair-West, 1971). The juxtaglomerular apparatus hypertrophies during Na deficiency in native marsupials (Blair-West et al., 1968).

The endocrine control of the marsupial kidney appears to be basically the same as in placentals.

E. THE ROLE OF THE GUT IN OSMOREGULATION

The gastrointestinal tract is the avenue through which most animals gain all their water and salts. Losses of fluids also occur via the gut, in the faeces.

The capacity of the marsupial gut to absorb water and salts has not been explored but there is no reason to suspect that it basically functions differently from that of placentals.

The macropodid marsupials have a large stomach which performs digestive functions similar to those of the rumen in the Ruminantia. It has been suggested by Macfarlane (1964) that the relatively large amounts of fluids that are sequestered in the gut of camels (equivalent to about 12–15% of their body weight), may act as a store that moderates the process of dehydration when water supplies are limited. In the euro, *M. robustus*, the contents of the large gut were found in the wild to weigh an amount which is equivalent to 10 to 15% of the body weight (Ealey et al., 1965). The gut fluids of kangaroos could thus play a similar role to that suggested for camels and act as a buffer that slows bodily dehydration. When the euros were dehydrated their stomach contents declined to very low levels but this change, at least partly, reflects an associated anorexia.

Water and salt loss occurs via the faeces. In placentals salt absorption occurs from the small intestine and colon and in the latter this process can be increased by aldosterone. No information is available as to whether the latter control mechanism exists in marsupials.

Faecal water loss varies a great deal in marsupials depending on their diet, whether they are in captivity or in the wild, and it reflects their state of bodily hydration. Three species of bandicoots with water available *ad libitum* in the laboratory, were found to lose water in their faeces which was equivalent to 12 to 22% of their total water loss each day (Hulbert and Dawson, 1974a). This quantity was of a magnitude equal to, or greater than, the water lost in the urine. When water was restricted the bandicoots' total faecal water losses, and the water content of the faeces, declined to less than half the amounts seen when water was available *ad libitum*. In the Kangaroo Island wallaby, *M. eugenii*, with water available *ad libitum* it was found that faecal water losses were nearly double those that occurred in the urine but, like in the bandicoots, when water was restricted the water losses in the faeces declined by almost 50% (Barker *et al.*, 1970). Quokkas, *S. brachyurus*, in the laboratory, have a very variable faecal water loss which can vary from about 5 to 40% of the total water loss each day (P. J. Bentley, personal observations). The loss of water in the faeces of marsupials may make a substantial contribution to the animals' water balance; it may be at least as great as that of the urine.

The mechanisms for reducing water excretion from the gut involve an absorption of the water across the gut wall. In placentals this process occurs principally in the colon. The water content of the faeces of Kangaroo Island wallabies and quokkas kept in captivity is high; about 3 g H_2O g dry wt^{-1} (Barker, 1971; P. J. Bentley, personal observations). Under field conditions, however, it is much less than this and in the dry season may be as low as 0·83 g H_2O g dry wt^{-1} in the Kangaroo Island wallaby and 1 g H_2O g dry wt^{-1} in the quokka (Barker, 1971; Bentley, 1955). These values suggest that the intestine of marsupials has an ability to adjust water reabsorption in response to the animal's needs.

F. EFFECTS OF REPRODUCTION ON WATER AND SALT METABOLISM

Like all mammals, marsupials suckle their young. The period of lactation is an extended one which in macropods may be more than six months in duration. During early lactation, the young, being very small, would make little demand on the water and salt resources of the mother. In the later part of the lactational period, however, the volumes of milk secreted are probably quite large.

The composition of the milk of marsupials has been shown to alter as lactation progresses (Lemon and Barker, 1967). The milk of the quokka initially has a high Na to K ratio but this gradually changes so that after about 150 days of lactation the K content exceeds that of the Na (Bentley and Shield, 1962).

A change in the nature of the salt demands on the mother must thus occur as lactation proceeds.

The water consumption of lactating marsupials would be expected to increase as the young one gets older and larger. During dry weather especially, lactating euros, *M. robustus*, in a wild population in north-west Australia were observed to drink more frequently than those which were not lactating (Ealey, 1967). More direct measurements of water turnover (with tritiated water) indicate the magnitude of the effect of lactation on the animals' water metabolism. In the short-nosed bandicoot, *I. macrourus*, lactation was found to increase water turnover by 36% (Hulbert and Dawson, 1972, 1974a). In the rock wallaby, *Petrogale inornata*, and the brush possum, *T. vulpecula*, it increased by 20% (Kennedy and Heinsohn, 1974). An interesting mechanism to reduce the overall loss of water in the milk has been described in the kangaroos *M. robustus* and *M. giganteus*, (Baverstock and Green, 1975). The mother, by consuming the urine and faeces of the pouch young, recycles about one-third of the water secreted in the milk.

It is uncertain what specific effect a limited water supply has on the survival of the pouch young under natural conditions. During drought there is a high rate of mortality in the pouch young of *M. rufa* (Newsome, 1965a). It is, however, not clear what the respective contributions of a lack of water and poor nutrition may be under these circumstances. Both conditions could result in a reduction or cessation of lactation and precipitate the death of the young.

The newborn marsupial only has a limited ability to regulate the composition of its body fluids and in some respects is analogous to late fetal stages of a placental mammal (Bentley and Shield, 1962). The newborn pouch young, or joey, of the quokka has a poorly developed kidney which, nevertheless, contains an active nephrogenic zone that is responsible for a rapid rate of renal growth and differentiation during the first 100 days after birth. During this period the young has a poor ability to concentrate its urine but this capacity increases after about 120 days. The content of neurohypophysial hormones in the pituitary also starts to rise at about the latter time. Homeothermy does not commence until the newborn quokka is about 100 days old (Shield, 1966). After that time VO_2 increases and the body temperature rises to a level which is slightly higher than that of the pouch. It appears, however, that the mother acts as a thermal sink to maintain the body temperature of the joey so that it may not need to expend substantial amounts of its own body water for evaporative cooling. Once the young starts to leave the pouch it will, presumably, lose water by evaporaton and as it is relatively small in size will be expected to become dehydrated more readily than the adults. The observed high mortality of older joeys during periods of drought may partly reflect this situation (Newsome, 1965a).

G. BEHAVIOUR AND OSMOREGULATION

Animals by adopting appropriate behaviour can considerably ameliorate the effects of a hostile environment on their water and salt balance.

1. *Selection of Food*

Little information is available about salt balance in marsupials but in areas where Na is deficient in the soil kangaroos are known to be attracted to salt-licks that are put out for cattle (Blair-West et al., 1968). Many native plants, especially those of the family Chenopodiaceae are halophytes and contain high concentrations of Na relative to K (Chippendale, 1965; Griffiths and Barber, 1966). Kangaroos eat such plants, especially, in some areas, bluebush (*Kochia*) and saltbush (*Atriplex*), but it is doubtful if this reflects any preference that involves satisfying a salt appetite

By preferentially grazing on the most green and succulent vegetation, marsupials that live in hot, dry areas may obtain additional water that promotes their survival and reduces their need for drinking water. It has been observed that kangaroos living in south-west Queensland, unlike sheep, seem to select the best blades of grass since even in the hottest weather "the stomach contents of kangaroos were green whereas in sheep at that time they were yellow with dried-off stalks" (Griffiths and Barker, 1966). Such green plants also usually have a higher protein content so that water may not be the only consideration in the choice.

The food preferences of the quokka, *S. brachyurus*, on Rottnest Island off the south-west coast of Australia, have been determined by identifying the remains of plant species that are present in their faeces (Storr, 1964). During the seasonal summer drought these small wallabies, especially those living in an area where there is no drinking water, ate large quantities of very succulent plants. While this choice apparently aids their water balance they suffer an associated malnutrition which may be due to the low protein content of this feed.

As described above, halophytic plants are also present in many parts of Australia and it has been observed that, in south-west Queensland, bluebush is eaten in large amounts by the red kangaroos but not so much by the grey kangaroos (Griffiths and Barker, 1966). Whether this is a positive choice or a necessity due to lack of other feed is not clear though it was noted that both species of macropodids as well as sheep "were driven to eat large amounts of Kochia" when other food was scarce. The osmotic effects of such food which contains substantial amounts of water but also a high salt concentration is unknown.

The congregation of red kangaroos around areas where green grass is available has been observed during a drought in central Australia (Newsome, 1965b, 1965c). The red kangaroos are nomadic and may travel considerable distances in search of nutritious food. It is considered likely that these kangaroos normally obtain sufficient water from succulent food but when this is scarce, and the environmental conditions are extremely hot, they are forced to seek drinking water.

Several families of Australian marsupials are carnivorous and insectivorous including the Dasyuridae and most of the Peramelidae (though some of the latter may eat vegetable food) (Troughton, 1943). The free and metabolic water supplied by such diets may vary but whether the animals select prey, for species with a high fat content, that may supply them with additional water, is unknown.

2. *Selection of Refuges*

By seeking refuge from the heat and solar radiation many marsupials reduce their evaporative water loss. Such behaviour must be of considerable importance for the survival of species that live in hot desert regions, especially during periods of drought. The nature of the refuges varies. Small animals have greater opportunities to find suitable havens than large ones and may resort to a nocturnal life and spend the daylight hours in burrows, hollow logs, small caves and crevices in rocks and nests of vegetation. The rabbit-eared bandicoot, *M. lagostis*, is the best adapted member of its family to desert life and is the only actively burrowing bandicoot. In the day it burrows deeply into the ground and so avoids the extreme desert heat (Troughton, 1943). The Dasyuridae also seek daily refuge in burrows and other protected places. Larger animals, however, generally cannot find such desirable refuges. Euros, *M. robustus*, live around rock outcrops and during the day seek refuge in the shade of overhanging rock ledges and in caves (Ealey, 1967a). The numbers of animals observed in such situations is much greater in hot weather and when water is scarce. The red kangaroos, *M. rufa*, are found in more open country and in the hot weather seek the shade of small trees (Newsome 1965b, 1965c). During periods of drought the red kangaroos in central Australia congregate (as we have seen) in areas near to where green grass is present and small, shady trees are available. They have been observed to assemble in woodlands which are close to the desired grazing areas, usually about 1·2 to 2·5 km away from dams and bores from which they can obtain drinking water. Following rains they disperse more widely.

Measurements have been made of the influx of solar radiation and the air temperature in typical refuges of the red kangaroo, in the shade of small trees, and of the euro, in small caves in a rock outcrop (Dawson and Denny, 1969). The sites chosen by the euros offer much greater protection from solar radiation than those occupied by the red kangaroos. Even the shade of trees, however, prevented the penetration of about 80% of the solar radiation into the sites where red kangaroos rest. The posture that the kangaroos adopt in these refuges may also influence their water loss. It has, for instance, been found (Dawson, 1972) that when the red kangaroos are in the shade they gain less solar radiation when they are lying than standing. In the former position, however, they may not be able to dissipate heat as readily nor would they be able to take advantage of any breezes that may aid in their cooling.

3. *Drinking Patterns*

From the available information given by biologists, naturalists, laymen and hearsay, it seems likely that many marsupials do not regularly require drinking water. Such reports even include animals that live in hot, arid areas. Populations of euros have been described that live in hot, dry areas in north-west Australia where there is no permanent water for them to drink (Ealey *et al.*, 1965; Ealey, 1967a). The tammar *M. eugenii*, lives on several islands off the Australian coast and on many of these fresh drinking water is not available for extended periods of time (Kinnear *et al.*, 1968). Red kangaroos living in central Australia may only need to drink during especially dry periods (Newsome, 1965c). Such herbivorous species thus do not appear to be especially dependent on drinking water except during prolonged periods of drought and in extremely hot weather. Carnivorous marsupials like the mulgara, *D. cristicauda*, apparently obtain adequate water from their diet (Schmidt-Nielsen and Newsome, 1962) but it is uncertain how widespread this phenomenon is among such species and whether or not it also applies under the most adverse conditions. Bandicoots kept in the laboratory on a commercial diet prepared for dogs (Good-O dog food) consume, and require, substantial amounts of drinking water (Hulbert and Dawson, 1974a). A population of an omnivorous species, the short-nosed bandicoot, *I. macrourus*, lives on an island, in a salt water lake on the New South Wales coast, where there is no fresh drinking water.

These bandicoots have a reduced water turnover compared to a mainland population which has drinking water available though they are still able to survive without it (Hulbert and Gordon, 1972). The area where they live is, however, not an arid one.

In the desert areas of Australia the natural sources of drinking water are usually pools that remain along the beds of dried-out water courses. If these sources dry out, kangaroos often dig into the soil there in order to gain subsurface water. Such natural water supplies have been augmented by dams, bores and wells which provide water or sheep and cattle. Kangaroos move into the vicinity of these manmade water supplies as the natural sources and grasses dry out and the weather becomes hotter (Newsome, 1965c; Ealey, 1967a).

Ealey (1967a) has recorded the drinking patterns of the euro in the arid regions of north-west Australia. These animals come in to drink more frequently during hot, dry weather, if they are lactating, and if rocky shelters are sparse in number. This behaviour presumably reflects the animal's increased need for water under these conditions. Individual euros show differences in the frequency with which they come in to drink. During an especially dry year over a 13-day period 2% of marked population of euros came in and drank seven times while 28% of the animals did not drink at all during that time. It has been suggested that the euro population may contain "frequent" and "infrequent" drinkers; the differences probably reflect behavioural, but also possibly physiological factors, that result in a lower water expenditure by the latter animals. During a seven-day period in the summer, 196 euros were observed to each drink an average of 2·3 l on each visit to the water troughs. This amount is equivalent to about 12% of their body weight per drink (Ealey et al., 1965) and may reflect the degree of their dehydration.

Brackish water is available in many areas of Australia and could be drunk by the native marsupials. In coastal areas, especially the offshore islands, several reports suggest that kangaroos and wallabies may drink sea water (R. Lukis and R. Ottaway quoted by Ealey et al., 1965; Kinnear et al., 1968). Such observations are difficult to confirm. Attempts have been made to see if marsupials can be pursuaded to drink sea water in captivity. The quokka, *S. brachyurus*, if dehydrated, will sometimes drink small amounts of sea water but it seems unlikely that this could result in any useful gain of water under natural conditions (Bentley, 1955). The tammar, *M. eugenii*, drinks sea water more readily and apparently can so prolong its survival in the absence of fresh water (Kinnear et al., 1968). About 25 days appears to be the limit of its survival under these conditions (Purohit, 1971). It seems likely that marsupials could, if necessary, drink brackish water in order to survive since (Table IV) their abilities to concentrate urinary salts are quite high. *Setonix brachyurus* in captivity on a dry diet can maintain a positive water balance when drinking 2·5% NaCl solutions (Bentley, 1955) and this ability is unlikely to be unique among marsupials.

H. EFFECTS OF THE AVAILABILITY OF WATER ON THE LIFE OF MARSUPIALS

Adequate supplies of water are necessary for the health and survival of animals and therefore it is not unexpected to observe that changes in its availability have profound effects on the life of marsupials. As Australia contains very large areas of arid territory where rain is meagre and often unpredictable, climatic effects on the native animals can be quite dramatic. The availability of water and the animals physiological condition are closely interwoven as dry conditions reduce the rate of growth, the nutritional value, and the water content of the feed. In addition, most animals reduce their food intake when they become dehydrated due to a lack of drinking water. When supplies of drinking water dry up, kangaroos, as described above, congregate in the area of any remaining water. The feed in the vicinity may then become overgrazed because the distances they can move for food are limited by the frequency that they may need to come in to drink. Such factors not only influence the lives of marsupials but also those of the cattle and sheep with which they share their habitat and with whom they compete for many types of feed.

The water and protein content of the plants on which animals graze reflect the local weather conditions so that during a period of drought the vegetation may supply little water and be of low nutritional value. Following rain, however, the plants rapidly revive. In the areas of north-west Australia inhabited by the euros studied by E. H. M. Ealey, the predominant feed grass is soft spinifex, *Triodia pungens*. Following a period of dry weather the water content of this grass may drop to 20% of the wet weight, and the protein present is 5% of the dry weight. After rain, however, the water content rapidly increases to 60% wet weight and the protein to 12% dry weight (Ealey *et al.*, 1965; Ealey and Main, 1967). Such changes in the composition of the plants can have dramatic effects on the survival and reproduction of the native animals.

1. *Hydration*

There are several physiological parameters that one can use to assess the animals state of hydration in nature. These include the measurement of tritiated water turnover, the concentrations of various plasma constituents, including Na, protein and haematocrit, as well as the urinary osmolality and volume.

Possums, *T. vulpecula* and rock-wallabies, *Petrogale inornata*, living in a subtropical woodland area in north Queensland had a similar water turnover at all seasons of the year (Kennedy and Heinsohn, 1974). Bandicoots, *I. macrourus*, that live in a temperate area on the New South Wales coast have a slightly lower water turnover in summer than winter suggesting that less water is available but this does not necessarily lead to any significant degree of dehydration (Hulbert and Gordon, 1972).

The quokka, *S. brachyurus*, on Rottnest Island experiences a dry summer and this is reflected by a reduction in urine volume and a rise in urinary osmolality (Bentley, 1955). As described earlier similar reductions in urine volume during the dry summer have also been recorded in *M. eugenii* on Kangaroo Island (Barker, 1971). Such changes in urine flow and concentration do not give any direct indication of the degree of dehydration but do show that mechanisms for water conservation are being brought into play.

The urinary concentrations in euros, *M. robustus*, living in the arid north-west of Australia were found to be similar throughout the year and as these levels were always high they suggest that an excess of water was rarely available (Ealey *et al.*, 1965). Red kangaroos and euros from dry areas in north-west New South Wales had a lower urinary concentration during the cooler times of the year and following rain (Dawson and Denny, 1969). The concentration of the urine of the red kangaroo was, however, always greater than that of the euro. In this study it was also found that the plasma Na concentrations in both species of kangaroos were elevated during a hot, dry period in December 1968 which probably indicates that they became dehydrated at this time. The euros in north-west Australia, however, had similar plasma Na concentrations at all times of the year (Ealey *et al.*, 1965). These levels had a mean of 153 mEq l^{-1}, which is similar to those of euros in captivity with water available *ad libitum*, 149 mEq l^{-1}. When captive euros were deprived of drinking water the plasma Na concentration rose to 169 mEq l^{-1}. It would thus appear that the wild euros in north-west Australia were not experiencing any severe dehydration. Many of them, however, were drinking regularly at a dam.

Four euros that were isolated in a natural area containing their normal food and a rocky shelter but without drinking water, lost weight but survived for nearly three months (Ealey *et al.*, 1965). Only one of these euros died before rain fell, though two more died shortly afterwards, probably as a result of malnutrition. Euros in the field without available drinking water clearly may suffer dehydration if the environmental conditions are severe but some may survive long enough to benefit from seasonal rains.

2. *Nutrition*

Macropodids may suffer from malnutrition during periods of drought. This was first clearly observed in the quokkas living on Rottnest Island which suffer a high mortality rate towards the end of summer. Their deaths can be correlated to an anaemia and a decrease in the plasma protein levels (Main *et al.*, 1959; Shield, 1959). These changes are thought to reflect change in the animals' diet (Storr, 1964; Tyndale-Biscoe, 1973; Barker *et al.*, 1974), especially the grazing of succulent plants which have a poor nutritional value. The consumption of such plants may be due to a lack of water or other food. The euros in north-west Australia also suffer from a seasonal anaemia (Ealey and Main, 1967) from which they recover after rain falls and plants with a higher protein content become available.

3. *Reproduction*

Mammals usually only come into breeding condition at a time when the availability of food and water augurs well for the survival of the young. In the drier or desert areas of Australia, this situation comes about following rains. Such opportunistic breeding is seen in a variety of Australian desert animals including birds, amphibians, fishes and marsupials. In the red kangaroo the signal to breed appears to be water, rather than the associated improved nutrition. It has thus been observed that they come into oestrus about two weeks after rain and as it takes 10 days for the Graafian follicles to develop, this time is too short to involve a change in the animals' nutrition (Sharman and Clarke, 1967; Tyndale-Biscoe, 1973). If conditions are moderately favourable euros and red kangaroos will breed throughout the year and even well into a period of drought (Newsome, 1965a; Sadlier, 1965). However, as a drought proceeds, and it may last for several years, the red kangaroos in central Australia have been observed to go into anoestrus and this appears to reflect a decline in their nutritional condition (Newsome, 1964, 1965a). Such fluctuations in breeding activity were not, however, observed in the euros and red kangaroos living in north-west Australia (Sadlier, 1965) but the drought conditions that they experienced were not, it seems, as severe as those in central Australia.

It has also been found that two populations of quokkas, one on Rottnest Island and the other at a proximate position on the mainland, experience different breeding seasons (Shield, 1965). The quokkas on the mainland, like domesticated animals from Rottnest Island, breed throughout the year while the island population experiences a seasonal anoestrus. It is thought that this situation reflects the seasonal semi-starvation that the animals from Rottnest Island undergo and which, as described above, may be related to the animals' need for water. The mainland quokkas live in an area where food and water is plentiful at all times of the year.

The macropodid marsupials have a delayed implantation of the blastocyst (more correctly called an embryonic diapause) which may remain in a quiescent state in the uterus for several months (Sharman, 1970). The ovum is usually fertilized during a post-partum oestrus but its development is delayed by the presence of a suckling young in the pouch. If this young is lost, then the further development of the blastocyst proceeds and another young is born about 30 days later. It has been suggested that the embryonic diapause may facilitate rapid successful reproduction at the end of a period of drought (Newsome, 1964, 1965a; Sadlier, 1965). When conditions become so severe that the kangaroo loses its suckling young, the dormant blastocyst will then start to develop. The resulting young will be born further into the drought period at which time, if anoestrus has not supervened, another ovum will be fertilized post partum. The new young will make few demands on the mother for a couple of months but if conditions remain unfavourable it may also die and the cycle is repeated. If, however, more favourable conditions intervene, the already developing joey may survive. Even if the drought is prolonged, so that anoestrus finally occurs, the development of the last blastocyst will project a young three months further into the dry period. During this time rain may fall so that it may then survive. The macropodid embryonic diapause may thus provide a mechanism that aids the rapid replenishment of the populations of kangaroos following prolonged periods of drought.

4. *Migration*

The migratory and nomadic habits of the red kangaroos in search of food and water have been described in an earlier section. Following rains, the population of red kangaroos in an area of 2600 m^2 in central Australia was observed to increase from about 4000 to 5000 which probably reflected an immigration from less favoured areas (Newsome, 1965b). Other macropodids, such as the euro and quokka, are more sedentary and this may result in more dramatic changes in their physiological condition during dry seasons.

5. *Abundance and Survival*

Macropodid marsupials it seems cannot live indefinitely without drinking water if the weather is very hot and their natural feed dries out. As described above euros may, however, survive two to three months of such harsh conditions. There are also, it will be recalled, reports of populations of euros living in remote desert areas and on islands where no drinking water is available. Such animals presumably eke out an existence by the efficient use of shelters and the choice of feed containing sufficient water for their needs. During prolonged droughts death is most probably only an indirect result of the absence of water and is due to starvation. The populations of macropodids have been seen to undergo spectacular changes in their size which can often be related to drought conditions. In north-west Australia, the droughts of 1934–35 and 1944–45, resulted in a marked depletion of both the populations of red kangaroos and euros (Ealey, 1967b). The euros subsequently recovered their numbers but the red kangaroos did not. The euro population, which now predominates in the area, was again decimated in 1953–54. In central Australia a prolonged drought from 1962–66 resulted in a halving of the population of red kangaroos in the area (Newsome *et al.*, 1967). These animals became emaciated under such conditions and the proximate cause of death was thought to be malnutrition and starvation. Indeed, as remarked upon by Ealey (1967b) the euros in north-west Australia during the drought of 1953–54 died "in thousands in caves and, strangly enough, around water holes". The populations of quokkas on Rottnest Island have also been observed to undergo spectacular "crashes" in their numbers during especially dry seasons, as in 1953–54; 1957–58 and 1961–62 (Tyndale-Biscoe, 1973).

III. MONOTREMES

A. DISTRIBUTION

The monotremes are represented by four extant species; *Tachyglossus aculeatus*, *Zagglossus bruijni*, *Z. bartoni* (the echnidas) and *Ornithorhynchus anatinus* (the platypus). *Tachyglossus aculeatus* (Troughton, 1943; Griffiths, 1968) occurs throughout Australia. It has been found in desert country in the Northern Territory, South Australia and Western Australia. *Zagglossus* is confined to New Guinea. Echidnas occupy areas that encompass a wide range of climatic conditions including tropical rain forests, dry arid deserts and areas where snow falls and the environmental temperature occasionally falls below zero. The platypus leads a semi-aquatic life living in burrows along the banks of streams and lakes in which they catch their food. They are confined to the eastern half of Australia extending from tropical northern Queensland to Tasmania. Only sporadic information is available about osmoregulation in the monotremes.

B. EVAPORATIVE WATER LOSSES

Monotremes are homeotherms but there have in the past been some reservations about this physiological character They have a much lower body temperature than other mammals, including marsupials; it is only about 30°C (Sutherland, 1897; Martin, 1903). Sutherland noticed a considerable range in the body temperature of his echidnas, *T. aculaeatus*. It could be as low as 22°C, at an air temperature of 14·4°C, while it was 36·6°C at an air temperature of 45°C. Martin found that both echidnas and platypus increased their metabolism at low environmental temperatures but, they had a poor ability to dissipate heat at high air temperatures. It was observed that if the body temperature rose to 37°C in the echidna or 35°C in the platypus, they soon died. These observations have been confirmed and extended in the echidna (Robinson, 1954; Schmidt-Nielsen et al., 1966) and the platypus (Robinson, 1954; Smyth, 1973). Such physiological characters have important effects on the animals water metabolism.

The relatively low body temperature in the monotremes has the consequence that they will tend to gain heat more rapidly from the environment than other mammals. As they also have a poor tolerance to an increased body temperature they can ill afford such an accumulation of heat in hot conditions. To compound the problem, monotremes have little ability to dissipate heat from their bodies. The evaporative water loss from the echidna is not significantly increased following exposure to environmental heat load (Robinson, 1954; Schmidt-Nielsen et al., 1966). The few observations that are available suggest that this situation also applies to the platypus (Robinson, 1954; Smyth, 1973). Neither of these monotremes increases its respiratory rate at high environmental temperatures and their skins appear to be only poorly supplied with sweat glands. Thus, they cannot readily utilize evaporation of water to prevent their body temperatures from rising in hot environments. As they also have a poor tolerance to elevate body temperatures it would appear that they must either reduce their heat production (which has not been shown in the echidna, Schmidt-Nielsen et al., 1966) or avoid such inequitable conditions. The latter solution appears to have been adopted in nature.

Evaporative water loss nevertheless occurs in monotremes and the results of Robinson (1954) suggest that this is substantially greater in the platypus than in the echidna. In the latter, this loss has been measured precisely and in an animal weighing 3 kg, at 25°C in dry air, it is about 50 ml day^{-1} while at 33·5°C it is 60 ml day^{-1} (Schmidt-Nielsen et al., 1966).

C. KIDNEY FUNCTION

Urinary water losses and concentrations have been measured in *T. aculeatus* (Bentley and Schmidt-Nielsen, 1967). It has been estimated under natural conditions at 25°C in dry air that the water loss in the urine is about 50% of the total loss. The echidna has a typical mammalian kidney with a loop of Henle and a medullary zone which is more than twice the width of the cortex. Urine concentrations as high as 2300 mosmol kg H_2O^{-1} have been recorded (Table IV) when food and water were withdrawn for short periods. The creatinine clearance in a normally hydrated echidna is somewhat lower than that of other mammals (Table III) and this decreased at low urine flows. This observation suggests that changes in the GFR, as well as renal tubular water reabsorption, may be contributing to control of urine flow as is frequently observed among non-mammals. Arginine-vasopressin and oxytocin have been identified in the echidna pituitary (Sawyer et al., 1960). It seems likely that the vasopressin functions as an antidiuretic hormone though its effect has not been investigated. The blood pressure of the platypus responds to the injection of vasopressin and although this is, no doubt, only a pharmacological effect it indicates that these animals can respond to such hormones (Feakes et al., 1950). Echidnas also display a mammalian type response to injected oxytocin which produces a release of milk during lactation (Griffiths, 1965a).

The adrenocortical steroid hormones, cortisol, corticosterone and aldosterone have been identified in both the echidna and the platypus (Table V). There are, however, some differences in their abilities to synthesize these hormones as corticosterone predominates over the other corticosteroids in the echidna but cortisol is more abundant in the platypus. The latter is the more usual situation in mammals but it is not unique (for instance corticosterone is predominant in rats and mice).

The role of the adrenal cortex has been studied in the echidna where it is somewhat remarkable compared to other mammals. Adrenalectomized echidnas maintain good physiological condition and experience no changes in their plasma or urinary levels of Na and K (McDonald and Augee, 1968). The adrenal cortex of the echidna is, however, clearly important in regulating their intermediary metabolism (Augee and McDonald, 1973). When echidnas are exposed to low environmental temperatures and food is in short supply, they enter a state of torpor, or hibernation when their body temperature declines to within a few degrees below the ambient. The adrenalectomized animals invariably went into torpor when they were exposed to cold, even when adequate food was available. The injection of cortisol or corticosterone prevented this response. The echidna appears to lack a mineralocorticoid response to corticosteroid hormones but their glucocorticoid effects are important in their adaptation to cold stress.

D. EFFECTS OF ENVIRONMENT AND BEHAVIOUR ON OSMOREGULATION

Little is known about osmoregulation in the platypus. As this animal occupies a habitat where water is always available and it has a poor ability to withstand high environmental temperatures it would not appear to be able to move far from such a habitat. Indeed, unlike the echidna, it is confined to the eastern part of Australia, possibly as a result of an inability to cross the central desert regions of the continent.

It is rather remarkable that the echidna which also has a poor ability to regulate its body temperature has such a wide distribution throughout such a climatically harsh continent as Australia. Its expenditure of water appears to be quite conservative as it can form a concentrated urine and does not utilize evaporation for temperature regulation. This strategy is reminiscent of that employed by small placental mammals which live in deserts, but these, because of their size, can readily obtain refuge in burrows and avoid excessive heat. The echidna is a larger beast but, it would seem, must also burrow to maintain some thermal stability. A prolonged study of their natural history has been made by S. J. J. Davies (quoted by Griffiths, 1968). Davies observed echidnas that lived near Mileura, close to the Murchison River (latitude 27° 30′) in Western Australia. The average annual rainfall in this region is an uncertain 25 cm and the environmental temperature ranges from 0 to 44°C. The echidnas studied by Davies lived in caves where the humidity was high and the air temperature showed little diurnal variation compared with the extremes in the open. When the conditions were favourable the echidnas emerged to feed, usually at night and after rain. The echidna has a remarkable ability to burrow into the ground and in the laboratory, when so entombed, can readily tolerate CO_2 levels as high as 5% (Bentley et al., 1967). They are, however, apparently not tunnellers and do not construct their own burrows but excavate depressions in the ground in which they cover themselves with dirt and debris. This habit could presumably afford them some protection against high air temperatures.

In captivity echidnas drink water. In many of the areas where they live, drinking water does not appear to be readily available but it seems likely that they get adequate water from their diet. In their natural habitat their food consists mainly of ants and termites (Griffiths, 1965,

1968; Griffiths and Simpson, 1966). They are especially partial to virgin queen ants (*Iridomyrmex detectus*) which contain 47% fat though they also thrive on less obese prey, containing about 70 to 80% water. It has been calculated that echidnas in their natural habitats should be able to maintain a positive water balance on this diet (Table VI). This estimate requires more direct confirmation under natural conditions.

TABLE VI

Water balance of the echidna *Tachyglossus aculeatus*

Intake (based on a diet of termites containing 77% water)	Water (ml day^{-1})
Free water	113
Metabolic water	
Fat	4
Protein (calc. from urinary urea)	5
Total	122

Output	Water (ml day^{-1})
Evaporation at 25°C	51
Water in	
Faeces	9
Urine	60
Total	120

These values are taken from various sources and represent the intake and output of water in an animal weighing 3 kg on a natural diet of termites and kept in a dry air at 25°C (from Bentley and Schmidt-Nielsen, 1967).

In dry air at 33·5°C evaporation would increase by 9 ml day^{-1}. In nature the air is never completely dry so that in their refuges evaporation would be expected to be less than the measurements above indicate.

IV. CONCLUSIONS

Marsupials occupy a great variety of habitats in Australia. Much of this continent is hot and dry and precipitation is either low or unpredictable. Many marsupials are, however, well adapted to life under such harsh conditions. Their strategies for survival are both physiological and behavioural and they do not seem to exhibit any adaptations that can be considered completely unique to the Marsupialia. This may reflect parallel evolution or the persistence of characters that were present in some common mammalian ancestor. Physiological mechanisms that are of special interest with respect to their osmoregulation, and which may influence their survival in arid habitats, include a lower body temperature and metabolic rate than is seen in most placental mammals, and an embryonic diapause. The monotremes have few species and offer little material for physiological diversity. The echidna has been most studied and apart from its phylogeny and oviparity is rather unusual for its low body temperature and rate of metabolism, very poor tolerance to high body temperatures, and its inability to dissipate heat effectively by evaporation. Despite these apparent paradoxes, it occupies a variety of habitats that include most of the hot, dry areas of Australia. The echidna is probably the most remarkable and unique of the native Australian mammals.

REFERENCES

Arnold, J. and Shield, J. (1970). Oxygen consumption and body temperature of the Chuditch (*Dasyurus geoffroii*) *J. Zool., Lond.* **160**, 391–404.

Augee, M. L. and McDonald, I. R. (1973). Role of the adrenal cortex in the adaptation of the monotreme *Tachyglossus aculeatus* to low environmental temperature. *J. Endocr.* **58**, 513–523.

Barker, S. (1971). Nitrogen and water excretion of wallabies: differences between field and laboratory findings. *Comp. Biochem. Physiol.* **38A**, 359–367.

Barker, S., Lintern, S. M. and Murphy, C. R. (1970). The effects of water restriction on urea retention and nitrogen excretion in the kangaroo island wallaby, *Protemnodon eugenii* (Desmarest). *Comp. Biochem. Physiol.* **34**, 883–893.

Barker, S., Glover, R., Jacobsen, P. and Kakulas, B. A. (1974). Seasonal anaemia in the rottnest quokka, *Setonix brachyurus* (Quoy and Gaimard) (Marsupialia; Macropodidae). *Comp. Biochem. Physiol.* **49A**, 147–157.

Bartholomew, G. A. and Hudson, J. W. (1962). Hibernation, estivation, temperature regulation, evaporative water loss, and heart rate of the pigmy possum, *Cercaertus nanus*. *Physiol. Zool.* **35**, 94–107.

Baverstock, P. and Green, B. (1975). Water recycling in lactation. *Science N.Y.* **187**, 657–658.

Bentley, P. J. (1953). Some aspects of the water metabolism and kidney histology of the marsupial *Setonyx brachyurus*. Honours B.Sc. Thesis, University of Western Australia.

Bentley, P. J. (1955). Some aspects of the water metabolism of an Australian marsupial *Setonyx brachyurus*. *J. Physiol. (Lond.)* **127**, 1–10.

Bentley, P. J. (1960). Evaporative water loss and temperature regulation in the marsupial *Setonyx brachyurus*. *Aust. J. exp. Biol. Med. Sci.* **38**, 301–306.

Bentley, P. J. and Schmidt-Nielsen, K. (1967). The role of the kidney in water balance of the echidna. *Comp. Biochem. Physiol.* **20**, 285–290.

Bentley, P. J. and Shield, J. W. (1962). Metabolism and kidney function in the pouch young of the macropod marsupial *Setonix brachyurus*. *J. Physiol. (Lond.)* **164**, 127–137.

Bentley, P. J., Herreid, C. F. and Schmidt-Nielsen, K. (1967). Respiration of a monotreme, the echidna, *Tachyglossus aculeatus*. *Am. J. Physiol.* **212**, 957–961.

Blair-West, J. R., Coghlan, J. P., Denton, D. A., Nelson, J. F., Orchard, E., Scoggins, B. A., Wright, R. D., Myers, K., Junqueira, C. L. (1968). Physiological, morphological and behavioural adaptation to a sodium-deficient environment by wild native Australian and introduced species of animals. *Nature, Lond.* **217**, 922–928.

Brown, G. D. (1969). Studies on marsupial nutrition. 6. The utilization of dietary urea by the euro or hill kangaroo, *Macropus robustus* (Gould). *Aust. J. Zool.* **17**, 187–194.

Buttle, J. M., Kirk, R. L. and Waring, H. (1952). The effect of complete adrenalectomy on the wallaby (*Setonyx brachyurus*). *J. Endocr.* **8**, 281–290.

Chippendale, G. (1962). Botanical examination of kangaroo stomach contents and cattle rumen contents. *Aust. J. Sci.* **25**, 21–22.

Coghlan, J. P. and Scoggins, B. A. (1967). The measurement of aldosterone, cortisol and corticosterone in the blood of the wombat (*Vombatus hirsutus* Perry) and the Kangaroo (*Macropus giganteus*). *J. Endocr.* **39**, 445–448.

Dawson, T. J. (1969). Temperature regulation and evaporative water loss in the brush-tailed possum *Trichosurus vulpecula*. *Comp. Biochem. Physiol.* **28**, 401–407.

Dawson, T. J. (1972). Likely effects of standing and lying on the radiation heat load experienced by a resting kangaroo on a summer day. *Aust. J. Zool.* **20**, 17–22.

Dawson, T. J. (1973). Thermoregulatory responses of the arid zone kangaroos, *Megaleia rufa* and *Macropus robustus*. *Comp. Biochem. Physiol.* **46A**, 153–169.

Dawson, T. J. and Brown, G. D. (1970). A comparison of the insulative and reflective properties of the fur of desert kangaroos. *Comp. Biochem. Physiol.* **37**, 23–38.

Dawson, T. J. and Denny, M. J. S. (1969a). Seasonal variation in the plasma and urine electrolyte concentration of the arid zone kangaroos *Megaleia rufa* and *Macropus robustus*. *Aust. J. Zool.* **17**, 777–784.

Dawson, T. J. and Denny, M. J. S. (1969b). A bioclimatological comparison of the summer day microenvironments of two species of arid-zone kangaroo. *Ecology* **53**, 328–332.

Dawson, T. J. and Hulbert, A. J. (1969). Standard energy metabolism of marsupials. *Nature, Lond.* **221**, 383.

Dawson, T. J., Denny, M. J. S. and Hulbert, A. J. (1969). Thermal balance of the macropodid marsupial *Macropus eugenii* Desmarest. *Comp. Biochem. Physiol.* **31**, 645–653.

Dawson, T. J., Robertshaw, D. and Taylor, C. R. (1974). Sweating in the kangaroo: a cooling mechanism during exercise, but not in the heat. *Am. J. Physiol.* **227**, 494–498.

Ealey, E. H. M. (1967a). Ecology of the euro, *Macropus robustus* (Gould), in north-western Australia. II. Behaviour, movements, and drinking patterns. *CSIRO Wildl. Res.* **12**, 27–51.

Ealey, E. H. M. (1967b). Ecology of the euro, *Macropus robustus* (Gould), in north-western Australia. I. The environment and changes in euro and sheep populations. *CSIRO Wildl. Res.* **12**, 9–25.

Ealey, E. H. M. and Main, A. R. (1967). Ecology of the euro, *Macropus robustus* (Gould) in north-western Australia. III. Seasonal changes in nutrition. *CSIRO Wildl. Res.* **12**, 53–65.

Ealey, E. H. M., Bentley, P. J. and Main, A. R. (1965). Studies on water metabolism of the hill kangaroo, *Macropus robustus* (Gould), in northwest Australia. *Ecology* **46**, 473–479.

Feakes, M. J., Hodgkin, E. P., Strahan, R. and Waring, H. (1950). The effect of posterior lobe pituitary extracts on the blood pressure of *Ornithorhynchus* (duck-billed platypus). *J. exp. Biol.* **27**, 50–58.

Ferguson, D. R. and Heller, H. (1965). Distribution of neurohypophysial hormones in mammals. *J. Physiol.* **180**, 846–863.

Frith, H. J. and Calaby, J. H. (1969). "Kangaroos." C. Hurst and Co., London.
Green, L. M. A. (1961). Sweat glands in the skin of the quokka of western Australia. *Aust. J. exp. Biol. Med. Sci.* **39**, 481–486.
Griffiths, M. (1965a). Rate of growth and intake of milk in a suckling echidna. *Comp. Biochem. Physiol.* **16**, 383–392.
Griffiths, M. (1965b). Digestion, growth and nitrogen balance in an egg-laying mammal, *Tachyglossus aculeatus* (Shaw). *Comp. Biochem. Physiol.* **14**, 357–375.
Griffiths, M. (1968). "Echidnas." Pergamon Press, New York.
Griffiths, M. and Barker, R. (1966). The plants eaten by sheep and by kangaroos grazing together in a paddock in south-western Queensland. *CSIRO Wildl. Res.* **11**, 145–167.
Griffiths, M. and Simpson, K. G. (1966). A seasonal feeding habit of spiny ant-eaters. *CSIRO Wildl. Res.* **11**, 137–143.
Gubbins, S. and Johnson, K. G. (1972). Evaporative water loss in the wallaby (*Protemnodon rufogrisea*). *J. Physiol. (Lond.)* **226**, 110–111P.
Haines, H., MacFarlane, W. V., Setchell, C. and Howard, B. (1974). Water turnover and pulmocutaneous evaporation of Australian desert dasyurids and murids. *Am. J. Physiol.* **227**, 958–963.
Hulbert, A. J. and Dawson, T. J. (1974a). Water metabolism in perameloid marsupials from different environments. *Comp. Biochem. Physiol.* **47A**, 617–633.
Hulbert, A. J. and Dawson, T. J. (1974b). Thermoregulation in perameloid marsupials from different environments. *Comp. Biochem. Physiol.* **47A**, 591–616.
Hulbert, A. J. and Dawson, T. J. (1974c). Standard metabolism and body temperature of perameloid marsupials from different environments. *Comp. Biochem. Physiol.* **47A**, 583–590.
Hulbert, A. J. and Gordon, G. (1972). Water metabolism of the bandicoot *Isoodon macrourus* Gould in the wild. *Comp. Biochem. Physiol.* **41A**, 27–34.
Hulbert, A. J. and Rose, R. W. (1972). Does the devil sweat? *Comp. Biochem. Physiol.* **43A**, 219–222.
Ilett, K. F. (1969). Corticosteroids in the adrenal venous and heart blood of the quokka, *Setonix brachyurus* (Marsupialia: Macropodidae). *Gen. Comp. Endocrinol.* **13**, 218–221.
Kennedy, P. M. and Heinsohn, G. E. (1974). Water metabolism of two marsupials—the brush-tailed possum, *Trichosurus vulpecula* and the rock-wallaby, *Petrogale inornata* in the wild. *Comp Biochem Physiol.* **47A**, 829–834.
Kennedy, P. M. and Macfarlane, W. V. (1971). Oxygen consumption and water turnover of the fat-tailed marsupials *Dasycercus cristicauda* and *Sminthopsis crassicaudata*. *Comp. Biochem. Physiol.* **40A**, 723–732.
Kinnear, J. E., Purohit, K. G. and Main, A. R. (1968). The ability of the tammar wallaby (*Macropus eugenii*, Marsupialia) to drink sea water. *Comp. Biochem. Physiol.* **25**, 761–782.
Lemon, M. and Barker, S. (1967). Changes in milk composition of the red kangaroo, *Megaleia rufa* (Desmarest), during lactation. *Aust. J. exp. Biol. Med. Sci.* **45**, 213–219.
Lintern, S. M. and Barker, S. (1969). Renal retention of urea in the kangaroo island wallaby, *Protemnodon eugenii* (Desmarest). *Aust. J. exp. Biol. Med. Sci.* **47**, 243–250.
McDonald, I. R. (1974). Adrenal insufficiency in the red kangaroo (*Megaleia rufa*, Desm.). *J. Eudocr.* **62**, 689–690.
McDonald, I. R. and Augee, M. L. (1968). Effects of bilateral adrenalectomy in the monotreme *Tachyglossus aculeatus*. *Comp. Biochem. Physiol.* **27**, 669–678.
Macfarlane, W. V. (1964). "Terrestrial Animals in Dry Heat: Ungulates." (J. Field, ed.) Sect. 4, Amer. Physiol. Soc. Williams and Wilkins, Baltimore, Maryland.

Macfarlane, W. V., Morris, R. J. H. and Howard, B. (1963). Turnover and distribution of water in desert camels, sheep, cattle and kangaroos. *Nature, Lond.* **197**, 270–271.

Macfarlane, W. V., Howard, B., Haines, H., Kennedy, P. J. and Sharpe, C. M. (1971). Hierarchy of water and energy turnover of desert mammals. *Nature, Lond.* **234**, 483–484.

MacMillen, R. E. and Nelson, J. E. (1969). Bioenergetics and body size in dasyurid marsupials. *Am. J. Physiol.* **217**, 1246–1251.

Main, A. R., Shield, J. W. and Waring, H. (1959). "Recent Studies on Marsupial Ecology." (A. Keast, R. L. Crocker and C. S. Christian, eds). W. Junk, Netherlands, 00–00.

Martin, C. J. (1903). Thermal adjustment and respiratory exchange in monotremes and marsupials—A study in the development of homoeothermism. *Phil. Trans. R. Soc. Lond. Ser* B **95**, 1–37.

Needham, A. D., Dawson, T. J. and Hales, J. R. S. (1974). Fore-limb blood flow and saliva spreading in the thermoregulation of the red kangaroo, *Megaleia rufa*. *Comp Biochem. Physiol.* **49A**, 555–565.

Newsome, A. E. (1964). Anoestrus in the red kangaroo *Megaleia rufa* (Desmarest). *Aust. J. Zool.* **12**, 9–17.

Newsome, A. E. (1965a). Reproduction in natural populations of the red kangaroo, *Megaleia rufa* (Desmarest), in Central Australia. *Aust. J. Zool.* **13**, 735–759.

Newsome, A. E. (1965b). The abundance of red kangaroos, *Megaleia rufa* (Desmarest), in Central Australia. *Aust. J. Zool.* **13**, 269–287.

Newsome, A. E. (1965c). The distribution of red kangaroos, *Megaleia rufa* (Desmarest), about sources of persistent food and water in Central Australia. *Aust. J. Zool.* **13**, 289–299.

Newsome, A. E., Stephens, D. R. and Shipway, A. K. (1967). Effect of a long drought on the abundance of red kangaroos in Central Australia. *CSIRO Wildl. Res.* **12**, 1–8.

Purohit, K. G. (1971). Absolute duration of survival of tammar wallaby (*Macropus eugenii*, Marsupialia) on sea water and dry food. *Comp. Biochem. Physiol.* **39A**, 473–481.

Reid, I. A. and McDonald, I. R. (1968a). Renal function in the marsupial *Trichosurus vulpecula*. *Comp. Biochem. Physiol.* **25**, 1071–1079.

Reid, I. A. and McDonald, I. R. (1968b). Bilateral adrenalectomy and steroid replacement in the marsupial *Trichosurus vulpecula*. *Comp. Biochem. Physiol.* **26**, 613–625.

Reid, I. A. and McDonald, I. R. (1969). The renin-angiotensin system in a marsupial (*Trichosurus vulpecula*). *J. Endocr.* **44**, 231–240.

Richmond, C. R., Langham, W. H. and Trujillo, T. T. (1962). Comparative metabolism of tritiated water by mammals. *J. Cell. Comp. Physiol.* **59**, 45–53.

Robinson, K. W. (1954). Heat tolerances of Australian monotremes and marsupials. *Aust. J. Biol. Sci.* **7**, 348–360.

Robinson, K. W. and Macfarlane, W. V. (1957). Plasma antidiuretic activity of marsupials during exposure to heat. *Endocrinology* **60**, 679–680.

Robinson, K. W. and Morrison, P. R. (1957). The reaction to hot atmospheres of various species of Australian marsupial and placental animals. *J. Cell. Comp. Physiol.* **49**, 455–478.

Sadleir, R. M. F. S. (1965). Reproduction in two species of kangaroo *Macropus robustus* and *Megaleia rufa* in the arid pilbara region of Western Australia. *Proc. zool. Soc. Lond.* **145**, 239–261.

Sawyer, W. H., Munsick, R. A. and Van Dyke, H. B. (1960). Pharmacological characteristics of neurohypophysial hormones from a marsupial (*Didelphis virginiana*) and a monotreme (*Tachyglossus* (*Echnida*) *aculeatus*). *Endocrinology* **67**, 137–138.

Schmidt-Nielsen, K. (1964). "Desert Animals: Physiological Problems of Heat and Water." Clarendon, Oxford, p. 201.

Schmidt-Nielsen, K. and Newsome, A. E. (1962). Water balance in the mulgara (*Dasycercus cristicauda*), a carnivorous desert marsupial. *Aust. J. Biol. Sci.* **15**, 683–689.

Schmidt-Nielsen, K., Dawson, T. J. and Crawford, Jr., E. C. (1966). Temperature regulation in the Echidna (*Tachyglossus aculeatus*). *J. Cell. Physiol.* **67**, 63–71.

Sharman, G. B. (1970). Reproductive physiology of marsupials. *Science N.Y.* **167**, 1221–1228.

Sharman, G. B. and Clark, M. J. (1967). Inhibition of ovulation by the corpus luteum in the red kangaroo, *Megaleia rufa*. *J. Reprod. Fert.* **14**, 129–137.

Shield, J. W. (1959). Rottnest field studies concerned with the Quokka. *J. R. Soc. Western Australia* **42**, 76–78.

Shield, J. (1965). A breeding season difference in two populations of the Australian macropod marsupial (*Setonix brachyurus*). *J. Mammal.* **45**, 616–625.

Shield, J. (1966). Oxygen consumption during pouch development of the macropod marsupial *Setonix brachyurus*. *J. Physiol.* (*Lond.*) **187**, 257–270.

Simpson, P. A. and Blair-West, J. R. (1971). Renin levels in the kangaroo, the wombat and other marsupial species. *J. Endocr.* **51**, 79–90.

Smith, H. (1951). "The Kidney. Structure and Function in Health and Disease." Oxford University Press, New York and Oxford.

Smyth, D. M. (1973). Temperature regulation in the platypus, *Ornithorhynchus anatinus* (Shaw). *Comp. Biochem. Physiol.* **45A**, 705–715.

Starr, G. M. (1964). Studies on marsupial nutrition. 4. Diet of the Quokka, *Setonix brachyurus* (Quoy and Gaimard) on Rottnest Island. *Aust. J. Biol. Sci.* **17**, 469–481.

Sutherland, A. (1897). The temperatures of reptiles, monotremes and marsupials. *Proc. R. Soc. Victoria* **9**, 57–67.

Troughton, E. (1943). "Furred Animals of Australia." Angus and Robertson, Sydney and London.

Tyndale-Biscoe, H. (1973). "Life of Marsupials." Elsevier, New York and Amsterdam.

Weiss, M. (1973). Biosynthesis of adrenocortical steroids by monotremes: Echidna (*Tachyglossus aculeatus*) and platypus (*Ornithorhynchus anatinus*). *J. Endocr.* **58**, 251–262.

Weiss, M. and McDonald, I. R. (1965). Corticosteroid secretion in the monotreme *Tachyglossus aculeatus*. *J. Endocr.* **33**, 203–210.

Weiss, M. and McDonald, I. R. (1966a). Adrenocortical secretion in the wombat, *Vombatus hirsutus* Perry. *J. Endocr.* **35**, 207–208.

Weiss, M. and McDonald, I. R. (1966b). Corticosteroid secretion in the Australian phalanger (*Trichosurus vulpecula*). *Gen. Comp. Endocr.* **7**, 345–351.

Weiss, M. and McDonald, I. R. (1967). Corticosteroid secretion in kangaroos (*Macropus canguru major* and *M.* (*Megaleia*) *rufus*). *J. Endocr.* **39**, 251–261.

Weiss, M. and Richards, P. G. (1970). Adrenal steroid secretion in the koala (*Phascolarctos cinereus*). *J. Endocr.* **48**, 145–146.

Weiss, M. and Richards, P. G. (1971). Adrenal steroid secretion in the Tasmanian devil (*Sarcophilus harisii*) and the eastern native cat (*Dasyurus viverrinus*). A comparison of adrenocortical activity of different Australian marsupials. *J. Endocr.* **49**, 261–275.

3. Rodents

H. J. FYHN
University of Oslo, Oslo, Norway

I.	Introduction	95
II.	Water Intake	97
	A. Food Water	97
	B. Drinking	98
	C. Water Turnover Rate	104
III.	Water Loss	105
	A. Evaporative Water Loss	105
	B. Sweat Glands	116
	C. Urinary Water Loss	117
	D. Faecal Water Loss	128
	E. Lactation	129
IV.	Conclusions	133
	References	135

I. INTRODUCTION

The order Rodentia is by far the largest of the mammalian orders both in terms of number of species and in number of specimens. Some 40% of the species of living mammals are rodents (Morris, 1965). Rodents are regarded as primitive mammals having a generalized rather than a specialized type of brain, skeleton and placentation. The fossil records are comparatively poor but points to a rapid radiation in Paleocene. The lack of fossils has caused taxonomists problems in the classification at the higher levels but commonly (Ellerman, 1940, 1941) rodents are subdivided into three main groups—the squirrel-like rodents (*Sciuromorpha*) with 366 species, the rat-like (*Myomorpha*) with 1183 species and the porcupine-like (*Hystricomorpha*) with 180 species (Morris 1965). Rodents have testified a high phylogenetic adaptability and are distributed all over the world's land areas having South America as their main region and Australia only sparsely populated. They are ubiquitous in the terrestrial environment having adapted to all available niches from polar to equatorial regions, from mountains to sea shore, and from deserts to swamps. Most rodents are terrestrial and fossorial but some are also arborial or semi-aquatic (Walker, 1964). An overwhelming majority of the rodents are small animals of 10 to 20 cm in body length and 50 to 500 g in body mass. Some few species, however, like the largest rodent, the capybara (*Hydrochoerus hydrochaeris*), and the beaver (*Castor fiber*), the North African crested porcupine (*Hystrix cristata*) and the mara (*Dolichotis patagonum*) have attained appreciable weights of 20 to 50 kg (Morris, 1965).

In spite of the wide distribution and diversity in habitats, the rodents are remarkably uniform in structural characters (Walker, 1964). A key character of the order is the sharp, curved and ever-growing incisors which are used in a chisel-like fashion during gnawing. The prevalent notion is that rodents are fundamentally herbivorous. This notion, however, is not substantiated by detailed, year-round food studies of rodents in the wild which more emphasize a diversity in the diet and a seasonal variation in the diet possibly depending upon the availability of foods in the habitat (Jameson, 1952; Landry, 1970; Batsli and Pitelka, 1971). Charges have thus been made against an allocation of the rodents to a certain trophic level as misrepresenting the situation occuring in the wild (Carleton, 1973). In turn, rodents have great ecological impact by being the main food source for carnivorous birds and mammals, and there is a well established dependency between the population size of the rodent prey and that of its predator. Rodents have been exploited as a source of food by man only to a very small extent, but in other respects these animals are of great importance to the well-being of man. Some rodents have been used extensively as test animals in the laboratory; the farming industry may benefit from their burrowing activity aerating the soil and from their decimation of insect populations; great damage, however, is done by rodents attacking and destroying agricultural crops and other goods, and by transmitting diseases to which man is susceptible.

Of mammals, the rodents have the most diverse distribution in terms of water availability in the habitat. They are present in the most arid deserts where drinking water is neither available nor is needed (Schmidt-Nielsen, 1964) as well as in swamps and along waterways where they live a semi-aquatic existence. Such diversity reflects a high capacity to compensate for adverse and varying external conditions. This chapter reviews the literature describing the means by which rodents master the varying osmotic conditions of their habitats.

Emphasis has been placed upon reviews and recent references rather than upon the historical development of our knowledge.

For a rodent to remain in an osmotic steady-state condition, obviously the gain and loss of water and salts must quantitatively be equal. Water and salts are gained through the intake of food and drink. Water will be lost by evaporation from the skin and respiratory tract, and together with salts from the kidneys, salivary glands, sweat glands, milk glands as well as in the faeces. Since, therefore, the body may gain and lose water and salts through several routes, a dealing with the osmotic regulation in rodents will be a multifactorial matter. In this chapter, the various factors will be discussed separately.

II. WATER INTAKE

Water may be gained by drinking, from free water in the food, and from oxidation water during metabolism of the food. The amount of water gained through each of these routes varies greatly from species to species and depend upon choice of food and availability of free water in the environment.

A. FOOD WATER

In arid regions, free water is seldom present and rodents inhabiting these regions must satisfy their water need by water in food. A carnivorous rodent like the grasshopper mouse, *Onychomus*, benefits from the high water content of their prey and remains in water balance on a diet of fresh liver without any drinking water (Schmidt-Nielsen and Haines, 1964; Whitford and Conley, 1971). The water content of plant material varies greatly with its succulence ranging from cacti with 80-90% water (Lee, 1963; Schmidt-Nielsen, 1964) to air-dried seeds containing 5-10% water (Schmidt-Nielsen, 1964).

Several rodents inhabiting deserts and arid regions have been maintained in the laboratory seemingly in a perfect condition when given a diet of dry seeds without any drinking water: e.g. *Dipodomys merriami* (Schmidt-Nielsen and Schmidt-Nielsen, 1951; Soholt, 1975; Vanjonack *et al.*, 1975), *D. deserti* (Vanjonack *et al.*, 1975), *Perognathus intermedius* (Bradley *et al.*, 1975), *P. californicus* (Tucker, 1965), *Jaculus jaculus* (Kirmiz, 1962; Katz, 1973), *Gerbillus gerbillus* and *Meriones crassus* (Katz, 1973) *Leggadina hermannsburgensis* (MacMillen *et al.*, 1972), *Cynomus ludovicianus* (Boice, 1972), *Peromyscus crinitus* (Abbott 1971), *Notomys alexis* (MacMillen and Lee, 1969), *Acomys russatus*

(Castel and Abraham, 1969; Shkolnik and Borut, 1969), *A. cahirinus* (Shkolnik and Borut, 1969), and *Mus musculus* (Haines and Schmidt-Nielsen, 1967; Castel and Abraham, 1969), These rodents have highly efficient mechanisms reducing the water losses to a minimum and rely heavily on oxidation water in their water balance (Schmidt-Nielsen and Schmidt-Nielsen, 1951; Haines et al., 1973a; Soholt, 1975). Other xeric rodents consume plant materials of higher succulence and meet most of their water requirements by the free water in the food (Schmidt-Nielsen, 1964; Chew, 1965; Raun, 1966). By food preference, therefore, rodents may adjust their water intake according to the need. The varied diet of the rodents (Landry, 1970) and the seasonal changes in food type composition (Whitaker, 1966; Bradford, 1974; Kritzman, 1974) can thus be regarded as behavioural adaptions to optimize not only the food and energy supply but also the water balance of the animal. Thus, consumption of seeds and arthropods by a deer mouse, *Peromyscus maniculatus*, in the wet season (January) equalled 81% and 16%, respectively, while in the dry season (July) this shifted to 27% and 58%, respectively (Jameson, 1952).

B. DRINKING

In mesic and moist environments, free water may generally be present and available for drinking. Few observations, however, have been made of rodents drinking in the field (Chew, 1965). In the laboratory when given access to water, most rodents drink readily, and the *ad libitum* water consumption has been measured in several rodents as summarized in Table I. Chew (1965) gives additional data from older literature. *Ad libitum* water consumption of animals given a dry feed can be used as a measure of the water requirements of the animals. It should, however, be recognized that various biotic and abiotic factors may influence the water requirement, and the values tabulated are specific for the conditions prevailing. A seasonal variation in water requirement is found for the antelope ground squirrel, *Ammospermophilus leucurus*, showing 50% higher daily water intake in April than in October (Maxson and Morton, 1974). Captivity and caging conditions may influence the water requirements. Thus, the spiny pocket mouse, *Liomys irroratus*, maintained in a terrarium with burrows drinks 40% of that of animals maintained in open cages (Hudson and Rummel, 1966). Female rodents may consume more water *ad libitum* than males (Thompson, 1971; Getz, 1968) but this is not always the case (Lee, 1963; Fisler, 1963).

A low relative humidity increases the minimum water requirement of the kangaroo rat, *Dipodomys agilis*, (Carpenter, 1966), and the *ad libitum* water intake of the meadow vole, *Microtus pennsylvanicus*, (Getz, 1963) but not of the prairie vole, *M. ochrogaster* (Getz, 1963), or of redback vole, *Clethrionomys gapperi* and white-footed mouse, *Peromyscus leucopus* (Getz, 1968). Population differences in water consumption exist within rodent species (Silverstein, 1961; Lee, 1963; Glen, 1970). The drinking activity of males of seven inbred strains of white mouse, *Mus musculus*, varies more than 100% from 0·16 to 0·34 ml H_2O g^{-1} day^{-1} (Silverstein, 1961). Boice (1972) suggested that rodents which in the wild exist without drinking water in captivity may develop a vital dependency upon drinking water. The antelope ground squirrel, *Ammospermophilus leucurus*, has an increased water intake during the first days after capture and the plasma osmolality of field animals equals that of animals deprived of water for 5 to 6 days in the laboratory (Hudson, 1962). This indicates that animals in the field are partly dehydrated, a conclusion also reached for several species of deer mice, *Peromyscus* (Fertig and Layne, 1963; Glen, 1970). A partial dehydration of the body may have adaptive value under conditions of water shortage. It may serve as a cue to water intake, and at the same time activate mechanisms which reduce water loss thus tending to restore the water balance of the body. A chronic partial dehydration has also been observed in man exposed to desert conditions (Rothstein *et al.*, 1947).

The daily water intake of mammals has been related to body weight by the power functions $I(ml\ H_2O\ g\ body\ wt^{-1}\ day^{-1}) = 0·24$ (g body wt)$^{-0·12}$ (Hudson, 1962 modified from Adolph, 1949) and $I(ml\ H_2O\ g\ body\ wt^{-1}\ day^{-1}) = 0·099$ (g body wt)$^{-0·097}$ (McManus, 1974 modified from Chew, 1965). Rodents may or may not comply with such relationships as is evident from Fig. 1. The compiled data correlate poorly with body weight. The level of water consumption of rodents seems better described by the relationship of Adolph (1943) than that of Chew (1965). With reference to the relationship of Adolph some rodents have water intakes corresponding to that expected from their weights (the moist *Napaeozapus insignis*, Brower and Cade, 1966; the mesic *Tamias striatus*, Forbes, 1967 and *Neotoma floridana*, Birney and Twomey, 1970; and the desert *Notomys cervinus*, MacMillen and Lee, 1969; Macfarlane *et al.*, 1971) while others deviate substantially. For instance, the arid *Perognathus fallax fallax* consumes 12% of the expected amount (MacMillen, 1964a), the arid *Cynomys ludovicianus* consumes 20% (Boice, 1972), the arid *Neotoma fuscipes macrotis* 270% (MacMillen, 1964a) and the moist *Clethrionomys gapperi* consumes from 200% to 520% of that expected from its weight (Brower and Cade, 1966; Getz, 1968; McManus, 1974).

There are large differences in water consumption of rodents from a

TABLE I

Ad libitum water consumption of various rodents given a dry laboratory diet

Species	Habitat	Body wt (g)	ml H$_2$O g^{-1} day^{-1}	T$_a$ °C	RH %	Reference
SCIUROMORPHA APLODONTIDAE						
Aplodontia rufa	moist	962	0.33	16–38	15–90	Nungesser and Pfeiffer (1965)
SCIURIDAE						
Cynomys ludovicianus	arid	1270	0.02	20–23	27–33	Boice (1972)
Citellus tridecemlineatus	arid	230	0.05	19–22	—	Armitage and Shulenberger (1972)
Ammospermophilus leucurus	desert	—	0.14	21–22	30–60	Maxson and Morton (1974)
A. leucurus	desert	99	0.13	22–25	40–60	Hudson (1962)
Tamias striatus	mesic	115	0.14	—	—	Forbes (1967)
Eutamias minutus	mesic	46	0.12	—	—	Forbes (1967)
E. merriami	arid	79	0.12	23–24	30–65	Wunder (1970)
E. quadrivittatus	arid	66	0.13	20–23	27–33	Boice and Witter (1970)
HETEROMYIDAE						
Perognathus f. fallax	arid	19	0.02	19–25	40–60	MacMillen (1964a)
Dipodomys ordii	arid	78	0.11	20–23	27–33	Boice (1972)
D. ordii	arid	83	0.11	20–23	27–33	Boice and Witter (1970)
D. agilis	mesic	60	0.28	21–25	25–60	Carpenter (1966)
D. a. agilis	arid	61	0.12	19–25	40–60	MacMillen (1964a)
Liomys salvani	arid	55	0.06	23–25	13–98	Hudson and Rummel (1966)
L. irroratus	arid	52	0.08	23–25	13–98	Hudson and Rummel (1966)
MYOMORPHA CRICETIDAE						
Reithrodontomys megalotis	moist	10	0.33	—	—	Fisler (1963)
R. m. limicola	moist	9	0.15	19–25	20–50	MacMillen (1964b)

Species	Habitat					Reference
R. m. longicaudus	mesic	10	0.10	19–25	20–50	MacMillen (1964b)
R. raviventris halicoetis	moist	13	0.19	20–26	76	Haines (1964)
R. raviventris	moist	10	0.22	—	—	Fisler (1963)
R. halicoetis	moist	12	0.23	—	—	Fisler (1963)
Peromyscus truei	arid	28	0.23	20–23	27–33	Boice (1972)
P. crinitus stephensi	arid	14	0.11	27	25–50	Abbott (1971)
P. floridanus	arid	35	0.11	23	26–80	Fertig and Layne (1963)
P. floridanus	arid	29	0.17	26–28	62–83	Glen (1970)
P. g. gossypinus	mesic	25	0.20	26–28	62–83	Glen (1970)
P. polinotus subgriseus	mesic	13	0.30	26–28	62–83	Glen (1970)
P. polinotus subgriseus	arid	15	0.19	26–28	62–83	Glen (1970)
P. leucopus	mesic	22	0.13	15	75	Getz (1968)
P. l. noveboracensis	mesic	20	0.12	25	40–85	Ernst (1968)
P. maniculatus gracilis	mesic	18	0.08	23	50–80	Brower and Cade (1966)
P. m. gambelii	arid	20	0.12	19–25	40–60	MacMillen (1964a)
P. californicus insignis	arid	34	0.13	19–25	40–60	MacMillen (1964a)
P. eremicus fraterculus	arid	19	0.13	19–25	40–60	MacMillen (1964a)
Onychomys leucogaster	arid	33	0.23	20–23	27–33	Boice (1972)
Neotoma micropus canescens	arid	294	0.15	20	15–45	Birney and Twomey (1970)
N. floridana osagensis	mesic	283	0.12	20	15–45	Birney and Twomey (1970)
N. lepida	desert	139	0.26	20–23	50–80	Lee (1963)
N. lepida	mesic	110	0.16	20–23	50–80	Lee (1963)
N. l. intermedia	arid	146	0.13	19–25	40–60	MacMillen (1964a)
N. fusipes macrotis	arid	188	0.35	19–25	40–60	MacMillen (1964a)
N. fusipes	arid	187	0.23	20–23	50–80	Lee (1963)
Cricetulus griceus	arid	29	0.12	21	50–60	Thompson (1971)
Mesocricetus auratus	arid	108	0.09	21	50–60	Thompson (1971)
Lemmus lemmus	moist	75	0.31	21	50	Mansfield (1970)
Clethrionomys gapperi	moist	24	0.86	30	—	McManus (1974)
C. gapperi	moist	23	0.32	15	75	Getz (1968)
C. gapperi	moist	24	0.38	23	50–60	Brower and Cade (1966)
Microtus agrestis	mesic	24	0.62	21	50–60	Thompson (1971)
M. pennsylvanicus	moist	29	0.28	25–27	5–95	Getz (1963)
M. p. pennsylvanicus	mesic	29	0.21	25	40–85	Ernst (1968)
M. c. californicus	mesic	42	0.27	18–23	—	Church (1966)

TABLE I (continued)

Species	Habitat	Body wt (g)	ml H_2O g^{-1} day^{-1}	T_a (°C)	RH (%)	Reference
M. ochrogaster	mesic	32	0·21	25–27	5–95	Getz (1963)
Lagurus lagurus	arid	20	0·08	21	50	Mansfield (1970)
Meriones unguiculatus	desert	59	0·10	19–28	50–86	McManus (1972a)
M. unguiculatus	desert	77	0·09	20–23	27–33	Boice and Witter (1970)
M. unguiculatus	desert	94	0·04	20–25	35–80	Winkelmann and Getz (1962)
Psammomys obesus	desert	170	0·06	25	50	Frenkel et al. (1972)
P. obesus	desert	—	0·23	23–28	30–55	Abdallah and Tawfik (1971)
MURIDAE						
Rattus norvegicus albino	moist	248	0·10	21	50–60	Thompson (1971)
Pseudomys australis	desert	50	0·14	25	30–40	Macfarlane et al. (1971)
Mus musculus albino, ♀	mesic	—	0·29	25	—	Smith and McManus (1975)
M. musculus albino	mesic	30	0·20	29	—	Haines and Shield (1971)
M. musculus albino	mesic	25	0·19	21	50–60	Thompson (1971)
M. musculus **feral**	mesic	21	0·15	18–22	47–53	Haines et al. (1973a)
M. musculus feral	mesic	22	0·19	29	—	Haines and Shield (1971)
M, musculus feral	mesic	20	0·20	30	30	Haines et al. (1969)
M. musculus feral	mesic	11	0·17	25	40–85	Ernst (1968)
Leggadina hermannsburgensis	arid	12	0·35	25	30–40	MacMillen et al. (1972)
Notomys alexis	desert	35	0·19	25	30–40	Macfarlane et al. (1971)
N. alexis	desert	29	0·27	25	30–40	MacMillen and Lee (1969)
N. alexis	desert	41	0·15	25	30–40	Macfarlane et al. (1971)
N. cervinus	desert	34	0·15	25	30–40	MacMillen and Lee (1969)
ZAPODIDAE						
Napaeozapus insignis	moist	21	0·16	23	50–60	Brower and Cade (1966)
HYSTRICOMORPHA						
CHINCHILLIDAE						
Chinchilla laniger	arid	500	0·06	22	43	McManus (1972b)

given habitat (Fig. 1) but, on the average, rodents show a decrease in water requirement with increasing aridity of their macro-environment (Table II). Taken that the variation in experimental conditions is similar for each group, this indicates that the water availability of the macro-environment is of importance in the selection of habitat of a rodent species. Detailed knowledge of the micro-environment, behaviour and food preference of a rodent is likely to give a different picture

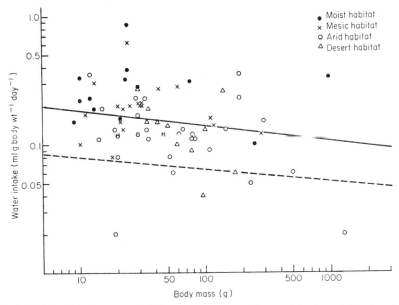

Fig. 1. Relationship between *ad libitum* water consumption and body weight of various rodents from different habitats. The lines describe the relationship between *ad libitum* water consumption and body weight of mammals generally as given by Hudson (1962, modified from Adolph, 1949) (———) and by McManus (1974, modified from Chew, 1965) (------).

of the availability of water to the animal in its niche than is apparent from the macro-environment. Such information may explain, at least partly, the large variation in water intakes of rodents from a given habitat. Thus, although living in desert and semi-desert areas, pack rats, *Neotoma* sp. and sand rat, *Psammomys obesus*, due to their preference for succulent plants (Raun, 1966; Birney and Twomey, 1970; Abdallah and Tawfik, 1971) and grasshopper mice, *Onychomys* sp., due to their carnivorous food habits (Schmidt-Nielsen and Haines, 1964; Whitford and Conley, 1971; Boice, 1972) would not experience the water shortage expected from their macro-environment. These rodents have an *ad libitum* water consumption well above that of other xeric rodents, (Table I).

TABLE II

Average *ad libitum* water intake and water turnover rate of rodent species from different habitats

Habitat	*Ad libitum* water intake		Water turnover rate	
	ml H_2O g^{-1} day^{-1}	Body wt (g)	ml H_2O g^{-1} day^{-1}	Body wt (g)
Moist	0·26 ± 0·04 (11)	9– 962	0·13, 0·20	297, 753
Mesic	0·20 ± 0·03 (16)	10– 283	0·22 ± 0·02 (9)	19–680
Arid	0·14 ± 0·02 (27)	12–1270	0·15 ± 0·03 (8)	21–412
Desert	0·17 ± 0·03 (6)	32– 170	0·08 ± 0·02 (9)	19–159

Data calculated from individual values in Tables I and III and given as mean ± s.e. above number of species in parenthesis or as individual values. For species with more than one set of data available the arithmetic mean value is used.

C. WATER TURNOVER RATE

The use of tritiated or deuterated water has offered a practical method for measuring the water turnover rate (WTR) in unrestricted rodents (Richmond *et al.*, 1962; Mullen, 1971; Yousef *et al.*, 1974). Tritiated water equilibrates with the body water in rodents (Holleman and Dieterich, 1973; Haines *et al.*, 1973b) and for an animal in water balance the daily WTR will be equal to the total water requirement per day (sum of water imbibed, water in food and oxidation water). Table III gives data on WTR of various rodents from different habitats. The WTR is calculated from the changing activity of tracer water with time in either body fluids (blood, urine) or pulmocutaneous evaporate. The two procedures would be expected to give equal WTRs in an animal maintained under standardized conditions. This has been found to be true for *Mus musculus, Peromyscus maniculatus* and *Notomys alexis* (Haines *et al.*, 1973a, 1973b). The WTR (ml day^{-1}) of rodents has been related to body weight by power functions with exponents of 0·80 (Richmond *et al.*, 1962), 0·78 (Holleman and Dieterich, 1973, related to lean body mass) and 0·82 (Yousef *et al.*, 1974).

From data on three species of the genus *Microtus* occupying a similar habitat, Holleman and Dieterich (1973) considered the value 0·69 as the best available estimate for an intraspecific exponent relating lean body mass to WTR (ml day^{-1}). The WTRs of Table III correlate poorly with body weight. Rodents from xeric environments tend to have a lower WTR than animals from mesic and moist habitats. On a mean basis the WTR correlates well with the ecological distribution of the animals (Table II) and agrees with the data on *ad libitum* water consumption. A relationship between WTR and habitat was suggested by Holleman and Dieterich (1973) and Yousef *et al.*, (1974). The limitations of the tritiated water method associated with extrapolating data on WTR obtained in laboratory experiments to the situation occurring in the field were discussed by Mullen (1970, 1971). In a combined field and laboratory study Bradford (1974) found that the arid adapted deer mouse, *Peromyscus truei*, had a daily WTR of 3·2 ml mouse^{-1} during the summer drought as compared to 6·7 ml mouse^{-1} and 0·8 ml mouse^{-1} in laboratory experiments on water *ad libitum* and on minimum water required for continued survival, respectively. Animal components, primarily insects, made up 55 65% of the summer diet of *P. truei* and probably were the main source of water for the animal. The WTR of *P. truei* in the field thus indicates the potentiality of the food as a water source of rodents during periods of water shortage.

III. WATER LOSS

A. EVAPORATIVE WATER LOSS

A continuous water loss occurs in terrestrial animals by evaporation from the skin and respiratory tract. The magnitude of these water losses depends upon ventilatory activity and upon physical factors of the micro-environment, i.e. air temperature, humidity and wind velocity (Ramsay, 1935; Leighly, 1937; Machin, 1964). Detailed and separate information on the pulmonary and cutaneous water loss would be necessary to prepare a comprehensive analysis of the evaporative water loss in rodents. Such data, however, are mostly lacking, a fact which may be related to the difficulties met with in the experimental procedures of such measurements in undisturbed rodents (Schmidt-Nielsen, 1964). From the available data however, it seems that large differences in the water loss via the two routes exist between rodent species. Thus, the desert-living Merriam's kangaroo rat, *Dipodomys merriami*, loses 16% of its total evaporative water loss from the skin at 26–27°C (Chew and Damman, 1961) while another desert inhabitant, the deer mouse, *Peromyscus maniculatus sonoriensis*, loses an equivalent of 46% from the

TABLE III

Tritiated water turnover rate in various rodents supplied with water and food *ad libitum*

Species	Habitat	Body wt (g)	ml H$_2$O g^{-1} day^{-1}	T$_a$ °C	Reference
SCIUROMORPHA					
SCIURIDAE					
Spermophilus lateralis	mesic	157	0·18	29	Yousef et al. (1974)[a]
S. tereticaudus	desert	117	0·12	29	Yousef et al. (1974)[a]
Ammospermophilus leucurus	desert	86	0·15	29	Yousef et al. (1974)[a]
Eutamias palmeri	mesic	69	0·31	29	Yousef et al. (1974)[a]
HETEROMYIDAE					
Dipodomys merriami	desert	34	0·03	29	Yousef et al. (1974)[a]
D. microps	desert	54	0·06	29	Yousef et al. (1974)[a]
D. deserti	desert	101	0·03	29	Yousef et al. (1974)[a]
D. deserti	desert	93	0·04	—	Richmond et al. (1962)
Perognathus formosus[b]	desert	19	0·03	25	Mullen (1970)
Liomys salvani	arid	55	0·11[c]	23–25	Hudson and Rummel (1966)
L. irroratus	arid	52	0·11[c]	23–25	Hudson and Rummel (1966)
MYOMORPHA					
CRICETIDAE					
Peromyscus truei	arid	21	0·32[c]	—	Bradford (1974)
P. maniculatus bardii	mesic	19	0·20	20	Holleman and Dieterich (1973)
Calomys ducilla	arid	27	0·12	20	Holleman and Dieterich (1973)
Neotoma lepida	arid	99	0·15	29	Yousef et al. (1974)[a]
Dicrostonyx groenlandicus	mesic	45	0·23	20	Holleman and Dieterich (1973)

Species	Habitat				Reference
Ondatra zibethica	semi-aquatic	753	0.20	20	Holleman and Dieterich (1973)
Microtus pennsylvanicus tananaensis	mesic	19	0.20	20	Holleman and Dieterich (1973)
M. oeconomus macfarlani	arid	45	0.23	20	Holleman and Dieterich (1973)
M. abbreviatus fisheri	mesic	50	0.29	20	Holleman and Dieterich (1973)
M. c. californicus	mesic	46	0.25	18–23	Church (1966)
Gerbillus gerbillus and G. campestre	desert	34	0.13[c]	26	Rouffignac and Morel (1966)
Meriones shawi	desert	159	0.08[c]	26	Rouffignac and Morel (1966)
M. unguiculatus	desert	80	0.03	20	Holleman and Dieterich (1973)
MURIDAE					
Rattus norvegicus albino	moist	371	0.16	20	Holleman and Dieterich (1973)
R. norvegicus albino	moist	221	0.11[a]	26	Rouffignac and Morel (1966)
R. norvegicus albino	moist	298	0.12	—	Richmond et al. (1962)
Mus musculus albino	mesic	22	0.34	—	Richmond et al. (1962)
M. musculus albino	mesic	32	0.18[c]	26	Rouffignac and Morel (1966)
M. musculus feral	mesic	21	0.22	20	Haines et al. (1973a)
Acomys cahirinus	arid	49	0.10	20	Holleman and Dieterich (1973)
HYSTRICORMORPHA					
CAVIIDAE					
Cavia porcellus	mesic	680	0.08	20	Holleman and Dieterich (1973)
CHINCHILLIDAE					
Chinchilla laniger	arid	412	0.06	20	Holleman and Dieterich (1973)

[a] Succulent vegetables as source of water.
[b] Deuterium—water.
[c] Body water content estimated to 65% (Holleman and Dieterich, 1973).
[d] Body water content estimated to 70% (Holleman and Dieterich, 1973).

skin at 27°C (Chew, 1955). Restrained specimens of *P. maniculatus*, at 29°C lose 56% of the total evaporative water loss from the skin when given water *ad libitum*, and 39% from the skin when given a restricted water ration of $\frac{1}{8}$ the *ad libitum* amount (Haines et al., 1969). A correction factor for the evaporation from head skin equalling 15% of the total evaporative water loss (Chew, 1955) has here been applied to the data of Haines et al. (1969). In restrained specimens of spiny mice, *Acomys cahirinus* and *A. russatus*, at 30°C, 60 to 70% of the total evaporative water loss was cutaneous (Shkolnik and Borut, 1969).

In comparison, the cutaneous evaporation of restrained white mouse *M. musculus*, was found to be below detection limits. In restrained specimens of feral and albino *M. musculus*, maintained on *ad libitum* water at 29°C, the ratios of the evaporative water loss between the respiratory tract-head area and the trunk area were found to be 65:35 and 63:37, respectively (Haines et al., 1969; Haines and Shield, 1971). By indirect means, the evaporative water loss has been partitioned into 37% from cutaneous sources and 63% from pulmonary sources in the California ground squirrel, *Spermophilus beecheyi*, at 28°C (Baudinette, 1972a), into 73% cutaneous and 27% pulmonary in male Syrian hamster, *Mesocricetus auratus*, and into 77% cutaneous and 23% pulmonary in the white rat, *Rattus norvegicus*, both at 30°C (Rodland and Hainsworth, 1974). The latter values are comparable with the findings of Tennent (1946) that 57% of the total evaporative water loss of the white rat at 23–25°C comes from the skin. Based on these few available data it thus seems that the evaporative water loss of rodents is shared equally between pulmonary and cutaneous water loss, possibly with the latter being the more important. An exception seems to be Merriam's kangaroo rat, *Dipodomys merriami*, which has a very low cutaneous water loss (Chew and Damman, 1961; Schmidt-Nielsen, 1964) attesting to the suitability of this genus for life under xeric conditions (Yousef et al., 1974).

Contrary to the scarcity of specific information on skin and lung perspiration, data on the total evaporative water loss, (the pulmocutaneous water loss), are available for numerous species of rodents (Table IV). The pulmocutaneous water loss is usually determined simultaneously with oxygen uptake measurements by using water vapour traps on the in- and outflowing air passages during open respirometry. The experiments have generally been carried out in "dry air", without further assessment of the actual humidity of the air. When such measurements are given, the relative humidity in the respirometer chamber at 28–30°C is found to be from 8 to 10% (MacMillen and Lee, 1970; Wunder, 1970; MacMillen et al., 1972) up to 15–30% (Shkolnik and Borut, 1969; Baudinette, 1972b). The dry air procedure, therefore, may not have standardized the experimental conditions completely causing some variability of the results.

A simple formula for predicting the relative humidity in the respirometer chamber from the amount of evaporative water added by the organism and the air flow rate through the chamber was presented by Lasiewski et al. (1966) and was shown to give excellent agreement between predicted and measured values. A method for controlling water vapour pressure of the air during open respirometry has been described (Ewing and Studier, 1973). The humidity of the air influences the pulmocutaneous water loss in rodents (Chew and Damman, 1961; Getz, 1968; Baudinette, 1972a) as shown in Fig. 2. Moreover, the dry air condition of the laboratory experiments has been criticized as abnormal even for desert rodents (Kirmiz, 1962; MacMillen and Lee, 1970) and may not be representative of the losses actually taking place in nature. Measurements of the microclimate in rodent burrows show the relative humidity to be well above that of ambient air ranging from 30 to 75% to almost saturation (Schmidt-Nielsen and Schmidt-Nielsen, 1950b; Kirmiz, 1962; McNab, 1966; Baudinette, 1972a). Schmid (1972) discussed how daily differences in water vapour pressure of the air would increase the pulmocutaneous water loss in nocturnal rodents relative to diurnal species.

Rates of pulmocutaneous water loss in Table IV are expressed as functions of body weight and oxygen consumption. The latter function allows for differences in metabolic rate of rodents of same body weight. Most of the data concern rodents in or close to the thermoneutral zone and all are well below the upper critical temperature. The values are thus unaffected by the strong increase in evaporative water loss which is found at high ambient temperatures. This increase is caused mainly by extensive salivation and body-licking (Breyen et al., 1973; Collins, 1973b; Rodland and Hainsworth, 1974; Bradley et al., 1975) but also by increased ventilation (Hudson and Deavers, 1973a, 1973b). Extensive salivation promoting evaporative heat loss seems to be a general thermoregulatory mechanism of rodents under stressfully high ambient temperatures.

In mammals, generally, pulmocutaneous water loss shows a correlation with body weight (Chew, 1965). The compiled data of Table IV, however, correlate poorly with body weight. When related to ecological distribution, there seems to be a correlation between aridity of the habitat and the metabolism-specific pulmocutaneous water loss (Table V). Rodents from xeric habitats have a significantly lower water loss than rodents from moist and mesic environments. Under conditions of water shortage it would be advantageous for the animal to keep the pulmocutaneous water loss at a minimum. Work by Haines and

TABLE IV

Pulmocutaneous water loss in various rodents supplied with water and food *ad libitum*

Species	Habitat	Body wt (g)	mg H_2O g^{-1} h^{-1}	mg H_2O cc O_2^{-1}	T_a (°C)	RH (%)	Reference
SCIUROMORPHA							
SCIURIDAE							
Spermophilus beecheyi	mesic	400–600	0.8	1.4	28	40	Baudinette (1972a)
S. lateralis	arid	164	1.6	1.0	29	0[a]	Yousef and Bradley (1971)
S. armatus	mesic	258	1.2	1.7	30	0	Hudson and Deavers (1973a)
S. beldingi	mesic	312	1.1	1.7	30	0	Hudson and Deavers (1973a)
S. spilsoma	mesic	302	1.4	2.5	30	0	Hudson and Deavers (1973a)
S. richardsoni	desert	128	1.1	1.8	30	0	Hudson and Deavers (1973a)
S. townsendi	mesic	251	1.4	2.3	30	0	Hudson and Deavers (1973a)
	arid	193	1.0	1.4	30	0	Hudson and Deavers (1973a)
Ammospermophilus leucurus	desert	99	1.6	1.4	30	0	Hudson (1962)
Eutamias merriami	arid	79	1.8	1.7	30	6–9	Wunder (1970)
HETEROMYIDAE							
Perognathus intermedius	desert	15	3.1	1.3	30	0	Bradley *et al.* (1975)
P. parvus		16	1.0	—		0	Anderson (1970)
P. penicillatus	mesic	18	1.6	0.9	29	0	Brower and Cade (1966)
P. californicus	arid	22	1.4	0.9	30	0	Tucker (1965)
P. baileye	desert	12–29	2.5	—	26–27	0	Chew and Dammann (1961)
P. baileyi	desert	25	—	0.5	28[b]	0	Schmidt-Nielsen and Schmidt-Nielsen (1950a)
Dipodomys microps	desert	62	1.3	1.1	30	0	Breyen *et al.* (1973)
D. merriami	desert	38	1.6	0.9	28	0	Carpenter (1966)
D. merriami	desert	30–37	1.5	—	26–27	0	Chew and Dammann (1961)
D. merriami	desert	36	—	0.5	28[b]	0	Schmidt-Nielsen and Schmidt-Nielsen (1950a)

Species	Habitat					Reference
D. agilis	mesic	60	1·5	1·1	28	0	Carpenter (1966)
D. spectabilis	desert	100	—	0·6	28[b]	0	Schmidt-Nielsen and Schmidt-Nielsen (1950a)
Liomys salvani	arid	55	0·6	0·9	28	0	Hudson and Rummel (1966)
L. irroratus	arid	52	0·6	1·0	28	0	Hudson and Rummel (1966)
MYOMORPHA							
CRICETIDAE							
Peromyscus maniculatus	arid	13–20	1·3	1·0	29	0	Haines et al. (1969)
P. m. sonoriensis	desert	16–18	3·7	0·9	27	6	Chew (1955)
P. m. gracilis	mesic	17	5·2	2·5	29	0	Brower and Cade (1966)
P. leucopus	mesic	22	4·5	—	30	0	Getz (1968)
P. crinitus	arid	22	—	0·5	28[b]	0	Schmidt-Nielsen and Schmidt-Nielsen (1950a)
Onychomys torridus	desert	19	5·8	3·7	30	0	Whitford and Conley (1971)
Neotoma lepida	desert	139	1·6	2·0	30	0	Lee (1963)
N. fuscipes	arid	187	1·3	1·6	30	0	Lee (1963)
Mesocricetus auratus, ♂	arid	80–120	1·8	—	30	<50	Rodland and Hainsworth (1974)
M. auratus	arid	95	—	0·6	28[b]	0	Schmidt-Nielsen and Schmidt-Nielsen (1950a)
Clethrionomys gapperi	moist	23	5·1	—	30	0	Getz (1968)
Microtus c. californicus	mesic	44	—	1·1	15–16	0	Church (1966)
M. pennsylvanicus	moist	23	4·8	1·6	28	0	Getz (1963)
M. ochrogaster	mesic	23	3·3	1·3	28	0	Getz (1963)
MURIDAE							
Rattus lutreolus	moist	109	1·4	1·6	28	0	Collins (1973a)
R. fuscipes		—	2·9	—	30	0	Collins (1973b)
R. rattus		132	1·5	1·3	28	0	Collins and Bradshaw (1973)
R. villosissimus	arid	187	1·1	1·7	28	0	
R. norvegicus albino	moist	140–220	1·1	0·9	30	≈75	Kirmiz (1962)
R. norvegicus albino	moist	102	—	0·9	28[b]	0	Schmidt-Nielsen and Schmidt-Nielsen (1950a)
Pseudomys australis	desert	50	3·7	1·4	25	0	Macfarlane et al. (1971)
Mus musculus albino	mesic	30	—	0·9	29	0	Haines and Shield (1971)

TABLE IV (continued)

Species	Habitat	Body wt (g)	mg H$_2$O g^{-1} h^{-1}	mg H$_2$O cc O$_2^{-1}$	(T$_a$) (°C)	RH (%)	Reference
M. musculus albino	mesic	27–44	—	0·7	29	0	Haines *et al.* (1969)
M. musculus albino	mesic	37	0·5	1·2	30	15–30	Shkolnik and Borut (1969)
M. musculus albino	mesic	9–24	3·9	—	26–27	0	Chew and Dammann (1961)
M. musculus albino	mesic	29	—	0·9	28^b	0	Schmidt-Nielsen and Schmidt-Nielsen (1950a)
M. musculus feral	mesic	21	5·4^d	—	18–22	47–53	Haines *et al.* (1973a)
M. musculus feral	mesic	22	—	0·9	29	0	Haines and Shield (1971)
M. musculus feral	mesic	18–34	—	0·9	29	0	Haines *et al.* (1969)
M. musculus feral	mesic	—	1·5	0·5	16–24	6	Fertig and Edmonds (1969)
M. musculus feral	mesic	27	—	0·6	28^b	0	Schmidt-Nielsen and Schmidt-Nielsen (1950a)
Leggadina hermannsburgensis	desert	12	4·0	1·2	28	10	MacMillen *et al.* (1972)
L. hermannsburgensis	desert	13	—	1·2	28	0	MacMillen and Lee (1967)
Notomys alexis	desert	33	2·6	1·4	28	23	Baudinette (1972b)
N. alexis	desert	35	3·6	1·2	25	0	Macfarlane *et al.* (1971)
N. alexis	desert	32	2·0	1·0	28	7	MacMillen and Lee (1970)
N. alexis	desert	29	—	0·9	28	0	MacMillen and Lee (1967)
N. cervinus	desert	41	3·2	1·2	25	0	Macfarlane *et al.* (1971)
N. cervinus	desert	34	2·7	0·8	28	6	MacMillen and Lee (1970)
N. cervinus	desert	35	—	0·8	28	0	MacMillen and Lee (1967)
Acomys russatus	desert	51	1·3	1·3	28	15–30	Shkolnik and Borut (1969)
A. cahirinus	desert	42	1·3	1·4	28	15–30	Shkolnik and Borut (1969)
ZAPODIDAE							
Napaeozapus insignis	moist	21	5·2	2·5	29	0	Brower and Cade (1966)
DIPODIDAE							
Jaculus orientalis	desert	163	0·7	0·8	30	≈75	Kirmiz (1962)

^{a} 0 indicates dried air.　　^{b} Personal communication in Hudson (1962).　　^{c} Six *P. baileyi* and two *P. intermedius*.　　^{a} Calculated.

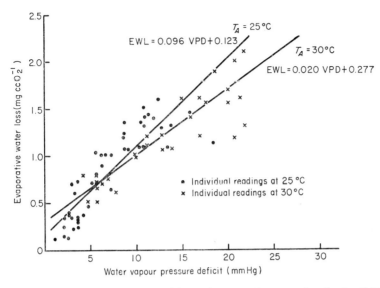

Fig. 2. Relation between ambient humidity and evaporative water loss in the California ground squirrel, *Spermophilus beecheyi*, at 25°C and 30°C (from Baudinette, 1972).

TABLE V

Average pulmocutaneous water loss of rodent species from different habitats

Habitat	Pulmocutaneous water loss		Body wt (g)
	mg H_2O g^{-1} h^{-1}	mg H_2O cc O_2^{-1}	
Moist	3·1 ± 0·9 (7)	1·6 ± 0·3 (5)	21–141
Mesic	2·3 ± 0·4 (11)	1·6 ± 0·2 (10)	18–500
Arid	1·3 ± 0·1 (10)	1·1 ± 0·1 (11)	17–193
Desert	2·1 ± 0·3 (16)	1·2 ± 0·1 (16)	12–163

Data calculated from individual values at 25–30°C in Table IV and given as mean ± s.e. above number of species in parenthesis. For species with more than one set of values available the arithmetic mean value is used.

collaborators has shown a significant reduction in this water loss upon prolonged water deprivation in feral and albino house mouse, *M. musculus*, and in deer mouse, *Peromyscus maniculatus*. When comparing animals maintained on water *ad libitum* with animals acclimated to a restricted water ration of $\frac{1}{8}$ of the *ad libitum* amount, *P. maniculatus* showed 45% reduction in the evaporative water loss (mg H_2O cc $O^{-1}{}_2$) (Haines et al., 1969), and feral and albino *M. musculus* had a reduction of 23% and 16–34%, respectively (Haines et al., 1969; Haines and Shield, 1971). Using an equal water restriction Haines et al. (1973a) found by indirect means that feral *M. musculus* had reduced the evaporative water loss (mg H_2O g^{-1} h^{-1}) by 70%. Similarly, when comparing specimens of Merriam's chipmunk, *Eutamias merriami*, which were water restricted or given water *ad libitum*, Wunder (1970) found a reduction in the evaporative water loss of 22% and 47% at 30°C and 38°C, respectively. The California vole, *Microtus c. californicus*, did not significantly reduce the pulmocutaneous water loss at 15–16°C (Church, 1966). Unfortunately, Church (1966) did not test the evaporation at higher temperatures. By partitioning the evaporative water loss between trunk and head-respiratory tract, Haines and collaborators (1969, 1971) invariably found that in water-restricted animals, the trunk showed the largest percentage reduction: 67% against 29% for *P. maniculatus*, 33% against 16–20% for feral *M. musculus*, and 44–58% against 0–16% for albino *M. musculus* (Haines et al., 1969; Haines and Shield, 1971). This is suggestive of a reduction in the transepidermal water loss upon water restriction. By applying a correction factor for the evaporation from the head skin of 30% of the total evaporative loss, Haines and Shield (1971) suggested that the reduction in total evaporative water loss could be due to changes in cutaneous evaporation only, while evaporation from the respiratory tract and eyes were constant. The mechanisms permitting the reduction apparently require several days or weeks to develop since prolonged and gradual water restriction was the causal factor (Haines et al., 1969).

The structural and physiological transformations resulting in reduced cutaneous evaporation upon water restriction are not clearly known (Haines et al., 1973a) but hypothetical explanations have been discussed (Haines and Shield, 1971). Especially, thickening of the stratum corneum has been suggested as causing the low cutaneous water loss in desert rodents (Sokolov, 1962; Rodland and Hainsworth, 1974). From measurements of the transepidermal water loss of various mammals including the white rat, *Rattus norvegicus*, Hattingh (1972a, 1972b) concluded that the thickness of stratum corneum *per se* was not the only factor in limiting the transepidermal water loss.

In a recent study on the transepidermal water loss in isolated rabbit

(*Lepus oryctolagus*) ear, Hattingh (1975) suggested that the amount of extracellular fluid in the skin is determining for the water loss. It appeared that the vascular system would regulate the amount of extracellular fluid in the skin according to the Starling hypothesis. Upon dehydration of the albino *M. musculus* the subcutaneous interstitial fluid pressure becomes increasingly more negative, probably caused by shifting of fluid from the extracellular space into plasma (Strømme et al., 1969). This could reduce the extracellular volume of the skin thus reducing the cutaneous evaporation according to the hypothesis of Hattingh (1975). In anaesthetized rats, *Rattus villosissimus* and *R. rattus*, cholinergic stimulation by intraperitoneal infusion of carbachol increased the evaporative water loss severalfold, probably by enhancing the cutaneous evaporation (Collins and Bradshaw, 1973).

It is of interest that vasopressin was found to decrease the transepidermal water loss in the rabbit ear (Hattingh, 1975) since an increase in the level of circulating vasopressin is a general response to water deprivation in mammals (Bentley, 1971) as in rodents (Castel and Abraham, 1969; El-Husseini and Haggag, 1974).

A low pulmonary water loss in rodents has been ascribed to a countercurrent heat and water exchanger in the nasal passageways (Schmidt-Nielsen and Schmidt-Nielsen, 1952; Jackson and Schmidt-Nielsen, 1964; Schmidt-Nielsen, 1969, 1972, 1975). The nasal countercurrent exchanger cools the expired air thereby reducing its water vapour content. This accomplishes conservation of body water as well as of body heat. There seems to be no difference between desert rodents (*Dipodomy, merriami* and *D. spectabilis*) and moist rodents (*R. norvegicus*) in this respect (Jackson and Schmidt-Nielsen, 1964). The quality is based upon the geometry of rodent nasal passageways which provide large surface area and short distance thus favouring heat and water exchange between the flowing air and the mucosa. The cooling effect is modified by changes in air temperature and humidity. In dry air the temperature of expired air of *Dipodomys* and *Rattus* was below the air temperature by 2-3°C (Jackson and Schmidt-Nielsen, 1964) and up to 10°C below air temperature in *D. spectabilis* at 35°C and 28% RH (Collins et al., 1971). Quantitating the water recovery due to the nasal countercurrent exchanger, Schmidt-Nielsen et al. (1970) showed a recovery of 54% for *D. merriami* at 30°C and 25% RH and of 83% at 15°C and 25% RH. Collins et al. (1971) developed and tested a steady state model for the nasal countercurrent exchanger in *D. spectabilis*. They found good consistency between predicted and measured values of expired air temperature and of net rate of water loss at different ambient conditions: predicted temperatures within 1·2°C of measured values, predicted water loss within 20% of measured values. The efficiency of water recovery (i.e. the ratio of water recovered on expiration to that evaporated on inspiration) was found to be uniformly high at about 70% over a wide range of ambient temperatures (10–36°C) and humidities (0–100% RW). Raab and Schmidt-Nielsen (1971) found that activity and the associated increase in air flow rates in the nose did not decrease the efficiency of the nasal countercurrent exchanger in *D. spectabilis* or in *R. norvegicus*. In the ground squirrel *Spermophilus beecheyi*, however, Baudinette (1972a) did not find a temperature depression of expired air at an ambient air temperature of 28°C. It should be of interest to verify the function of the nasal countercurrent heat and water exchanger in other rodents. Also it would be pertinent to investigate the effect of body dehydration on the efficiency of the water recovery in the nasal passageways.

It has been suggested (Riad, 1960; Babero et al., 1973) that the absence of mucus secreting goblet cells in the trachea of the kangaroo rat, *D. merriami*, and the scarcity of such cells in other desert rodents as compared with the white rat, may be an adaptation for water conservation. The mechanism for this adaptation was, however, not elaborated upon.

Colonial living is a common feature of fossorial rodents (Walker, 1964). The effect of this grouping on the water metabolism of the animals has been of little concern. The Australian hopping mouse, *Notomys alexis*, has colonial nesting habits and Baudinette (1972b) found the pulmocutaneous water loss (mg H_2O cc O_2^{-1}) to be 25% less in grouped (groups of 4) than in single animals at 28°C and 23% RH. Social aggregation, therefore, appears to have possible significance in water savings of the animals.

B. SWEAT GLANDS

Sweat glands are absent from the general body surface of rodents and present only in the restricted areas of the hairless footpads (Sokolov, 1962; Jenkinson, 1970, 1973; Collins, 1973b; Collins and Bradshaw, 1973). The function of these sweat glands has not been ascertained but it is unlikely that they are thermoregulatory organs (Jenkinson, 1973).

From their highly restricted distribution, it can be assumed that they play an insignificant role in the salt and water balance of the animal. Considering the large and unfavourable surface to body volume ratio of most rodents, Schmidt-Nielsen (1964) calculated that the water loss associated with a profuse sweating makes evaporative cooling an impossible thermoregulatory mechanism in rodents except for brief emergency situations. The lack of sweat glands in rodents is in line with this contention and may be regarded as an anatomical adaptation to water conservation.

C. URINARY WATER LOSS

1. *Kidney Structure*

The rodent kidneys are typically bean-shaped organs with a smooth surface and a size varying from 70 × 43 × 29 mm in the North African crested porcupine, *Hystrix cristata*, to 9 × 5·5 × 4·5 mm in the house mouse, *M. musculus* (Sperber, 1944). The kidneys are primitively simple having one renal lobe which can be differentiated into a cortical layer, a subcortical layer, an outer medullary zone and an inner medullary zone. The latter zone forms the papilla which extends into the renal pelvis.

In some species, especially xeric forms, the papilla may be long and project into the ureter as in *Dipodomys*, *Meriones*, *Psammomys*, *Notomys* and *Leggadina* (Sperber, 1944; Abdallah and Tawfik, 1969; MacMillen and Lee, 1969; MacMillen *et al.*, 1972). In other species, especially moist and semiaquatic forms, the papilla may be absent as in *Hydrochoerus*, *Hydromys*, *Hystrix*, *Myocastor* and *Aplodontia* (Sperber, 1944; Pfeiffer *et al.*, 1960). These latter rodents have a crest kidney. In some species there are small ridges of outer medullary zone tissue at the basis of the papilla called secondary pyramids (Pfeiffer, 1968, 1970; Zahn, 1968). The renal pelvis may be an uncomplicated, slightly expanded ureteral ending as in *Aplodontia rufa*, or have elaborate folds called specialized fornices that reach deep into the outer zone of the medulla forming the secondary pyramids as in *R. norvegicus* (Pfeiffer, 1968, 1970). The presence of secondary pyramids and specialized fornices seems to be related to the concentrating power of the kidney, possibly by enhancing recycling of urea within the medulla (Zahn, 1968; Pfeiffer, 1970).

Microscopically the rodent kidney conforms with the general description of the mammalian kidney as given by Rouiller (1969). The ultrastructure of rodent nephrons has been studied by scanning and transmission electron microscopy e.g. nephridial tubules (Latta *et al.*, 1967; Ericsson and Trump, 1969; Zimny *et al.*, 1971; Bulger *et al.*, 1974; Dieterich *et al.*, 1975), glomerulus (Simon and Chatelanat, 1969; Arakawa, 1970; Spinelli *et al.*, 1972; Andrews and Porter, 1974), basement membrane (Zimny, 1968; Sachot *et al.*, 1975) and papilla and pelvis surface (Bulger *et al.*, 1974). For the present purpose it suffices to give a short description of the functional microanatomy of a typical rodent kidney as emerging from the comprehensive studies on the rat kidney by Kriz (Kriz, 1967; Kriz *et al.*, 1972; Kriz and Lever,

1969). Figure 3 gives a schematic representation of the arrangement of vascular and nephridial structures in the rat kidney.

The cortex contains renal corpuscles, convoluted parts of the proximal and distal tubules, connecting portions of the collecting ducts and peritubular capillaries entangling the nephridial tubules. It seems that a countercurrent flow system exists between the nephridial tubules and the peritubular capillaries in the superficial renal cortex of the rat (Steinhausen et al., 1970a, 1970b).

The subcortical zone contains the straight parts of the proximal and distal tubules, collecting ducts, peritubular capillaries and the base of arterial and venous vasa recta derived from the efferent arterioles of juxtamedullary glomeruli.

The medulla contains short and long loops of Henle, collecting ducts, and vasa recta loops. The structure of the medulla is highly organized giving a striated appearance in saggital sections. The striae are formed by regularly repeating vascular bundles which traverse like inverted cones almost to the tip of the papilla. The vascular bundles consist of a central core of alternating arterial and venous vasa recta and a peripheral ring of venous vasa recta and descending thin limbs of the loop of Henle. Surrounding each vascular bundle is an area of collecting ducts and ascending limbs of the loop of Henle. The structural arrangement of the vascular bundles is suggestive of an efficient countercurrent exchange function. The thin ascending limbs of long loops of Henle are in juxtaposition to collecting ducts in the inner zone of the medulla thus being suited for countercurrent exchange. In the outer zone of medulla the thick ascending limb of the loop of Henle and the collecting ducts seem mostly separated by capillaries.

Comparisons with other rodents (*Meriones shawi*, *Mus musculus*, *Mesocricetus auratus*, *Ondatra zibethica*) show fundamentally the same structural arrangement of the medullary zones although quantitative differences exist as to the area and regularity of the vascular bundles and of the association between collecting ducts and ascending limbs of the loop of Henle (Kriz, 1970; Kriz and Koepsell, 1974). The highest regularity in tubular pattern was found in the desert species, *Meriones shawi*, and may probably be reflected in the efficient kidney function of this animal. It may be pertinent to recall at this point that rodents with a crest kidney lack an inner medullary zone.

A model for the medullary circulation in the chinchilla, *Chinchilla laniger*, was proposed by Prong et al. (1969). The model represents three major flow routes in series with the juxtamedullary glomeruli: through capillaries of the convoluted tubules, through capillaries at the junction between the outer and inner medullary zone (the frizzled zone) and through capillaries at the tip of the papilla.

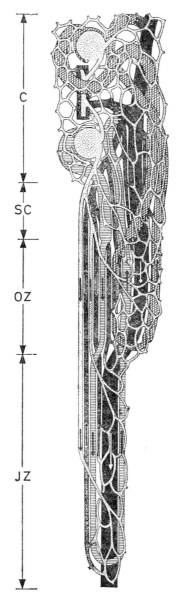

Fig. 3. The arrangement of nephridial and vascular structures in the rat kidney. The following structures are drawn: a cortical nephron with short loop of Henle (ruled rhombicly), a juxtamedullary nephron with long loop of Henle (hatched), a collecting duct (black), arterial vessels and capillaries (dotted loosely) and venous vessels (dotted compactly). C, cortex; SC, subcortical zone; OZ, outer zone of medulla; IZ, inner zone of medulla (from Kriz, 1967).

2. Kidney Function

The kidneys homeostatically control the composition and volume of the body fluids by adjusting the excretion of water and solutes according to the conditions of the blood plasma. The elaboration of urine results from an integrated co-operation between glomerular ultrafiltration, tubular reabsorption and tubular secretion and the resulting urine varies greatly in concentration and volume depending upon the water balance of the animal. A major portion of our knowledge on mammalian kidney function has been derived from experiments on laboratory rodents, especially the white rat, and the description of kidney function as found in general physiology textbooks, therefore, is characteristic of the rodent kidney. The physiology of the mammalian kidney has recently been treated in excellent reviews and monographs (e.g. Thurau, 1974; Sullivan, 1974; Valtin, 1973; Riegel, 1972) and will not be dealt with here. However, since the capacity of the rodent kidney to elaborate urine ranging in osmolality from a low 60 mosmol (*Chinchilla laniger*, Prong et al., 1969) to a high 9374 mosmol (*Notomys alexis*, MacMillen and Lee, 1969) has a direct bearing upon the osmotic regulation of these animals, the current hypotheses concerning the mechanism of urine concentration and dilution will briefly be discussed. Recent work on this subject has been reviewed (Jamison, 1974; Morel and Rouffignac, 1973; Marsh, 1971; Dicker, 1970).

a. *Mechanisms of urine concentration.* It is now generally recognized that the final concentration of the urine is accomplished by passive reabsorption of water from the fluid in the collecting ducts while traversing the hyperosmotic medulla. Water absorption is promoted by presence of the antidiuretic hormone (ADH) which increases the water permeability of the collecting ducts (mode of action of ADH reviewed by Dicker, 1970 and Grantham, 1974). Moreover, it is agreed that the hyperosmolality of the medulla results from a countercurrent multiplier system within the medulla. However, the exact mechanism of the multiplier system, and the site of energy for the osmotic work in the thin ascending limb of the loop of Henle are still matters of dispute (Jamison, 1974).

It is not fully resolved whether the concentration of tubular fluid in the thin descending limb of the loop of Henle depends upon solute addition as advocated by Rouffignac (Rouffignac and Morel, 1969; Rouffignac, 1972; Rouffignac et al., 1973) from work on the fat sand rat, *Psammomys obesus*, or whether it depends upon water extraction as suggested by Kokko (1970, 1972) working on isolated loops of Henle from the rabbit. This discrepancy may result from species specific differences but possibly also from differences in physiological conditions.

Jamison et al. (1973) found water extraction to be the dominant mode of concentration of the tubular fluid of the descending limb of the loop of Henle in water diuretic white rats, while solute addition and water extraction contributed equally to the concentration of this fluid in antidiuretic rats. However, Pennell et al. (1974) found that fluid within the descending limb of the loop of Henle in white rats became concentrated 60 to 67% by water extraction and 33 to 40% by solute entry regardless of the diuretic state of the animals.

Active transport of NaCl seems to take place in the thick ascending limb of the loop of Henle but the mode of transport, whether active or passive, in the thin ascending limb has not been settled (Valtin, 1973). Probably is Cl^-, not Na^+, the principal ion actively transported in the thick ascending limb of the loop of Henle (Burg and Green, 1973; Rocha and Kokko, 1973). Recently, models which account for the hyperosmolality of the inner medulla without involving active transport by the thin loops of Henle have been developed (Kokko and Rector, 1972; Stephenson, 1972). In these models, the outward transport of NaCl in excess of water in the thin ascending limb of the loop of Henle may occur passively down a concentration gradient and be effected by an opposite concentration gradient for urea. The concentration gradients are established by countercurrent multiplication in a vasa recta-loop of Henle-collecting duct complex as proposed by Kriz and Lever (1969), and by differences in permeability for water, salt and urea of the various structures involved. Urea is effectively being recycled within the medulla. An important function is ascribed to urea: its concentration in the distal tubules and collecting ducts provides energy to drive NaCl out of the thin ascending limbs of the loop of Henle. There is strong evidence of an active transport of urea out of the collecting ducts in protein-depleted white rats (Clapp, 1966; 1970; Truniger and Schmidt-Nielsen, 1970; Heuer et al., 1974). In the model of Stephenson (1972) the possibility that active transport of NaCl may occur in the thin ascending limb is not excluded but may operate parallel with the passive Na^+ transport. These models for the hyperosmolality of the papilla interstitial fluid have many attractive features but their exactitude is still open to discussion (Rouffignac et al., 1973; Jamison, 1974; Pennell et al., 1975).

b. *Urine osmolality.* The ability of a rodent to remain in osmotic balance under conditions differing widely in water and salt availability depends upon the capacity of the kidneys to dilute and concentrate the urine. Much interest has been centred upon the concentrating power of rodent kidneys (Table VI). Small, desert-living rodents have been found to concentrate the urine to values unbeatable by other mammals. In *Notomys alexis* a maximal urine osmolality of 9370 mosmol with a

TABLE VI

Urine osmolality (mosmol) and urine urea concentration (mM) of various rodents during dehydrating conditions

Species	Habitat	Osmolality		Urea concentration		Water ration	Reference
		mean	max	mean	max		
SCIUROMORPHA							
APLODONTIDAE							
Aplodontia rufa	moist	550	—	—	480	deprived	Schmidt-Nielsen and Pfeiffer (1970)
A. rufa	moist	770	—	—	—	deprived	Nungesser and Pfeiffer (1965)
A. rufa	moist	411	495	—	—	deprived	Dicker and Eggleton (1964)
A. rufa	moist	456	—	—	—	ADH inject.	House et al. (1963)
SCIURIDAE							
Spermophilus columbianus	arid	1639	2105	806	1740	deprived	Passmore et al. (1975)
S. columbianus	arid	1470	2792	695	1181	deprived	Moy et al. (1972)
Ammospermophilus leucurus	desert	3600	4055	—	—	minimum	Maxson and Morton (1974)
A. leucurus	desert	3730	—	—	—	deprived	Hudson (1962)
HETEROMYIDAE							
Dipodomys merriami	desert	3800	5800	1590[a]	2570[a]	deprived	Soholt (1975)
D. merriami	desert	4800	6350	—	—	deprived	Carpenter (1966)
D. agilis	mesic	4100	5150	—	—	deprived	Carpenter (1966)
Liomys salvani	arid	4000	4720	—	—	deprived	Hudson and Rummel (1966)
L. irroratus	arid	3580	4280	—	—	deprived	Hudson and Rummel (1966)

Species	Habitat						Condition	Reference
CASTORIDAE								
Castor canadensis	semi-aquatic	—	—	537	—	190	deprived	Schmidt-Nielsen and O'Dell (1961)
C. canadensis	semi-aquatic	—	—	—	—	375	deprived	Schmidt-Nielsen *et al.* (1961)
MYOMORPHA								
CRICETIDAE								
Reithrodontomys raviventris halicoetis	salt marsh	3170	4180	1990	—	2590	restricted	Haines (1964)
Peromyscus truei	arid	4250	4750	—	—	—	minimum	Bradford (1974)
P. crinitus stephensi	arid	3150	3430	1958	—	2214	deprived	Abbott (1971)
P. maniculatus bairdii	arid	6111	—	—	—	—	NaCl inject.	Heisinger *et al.* (1973)
P. m. nebrascensis	arid	5465	—	—	—	—	deprived	Heisinger *et al.* (1973)
P. leucopus novaboracensis	mesic	3838	—	—	—	—	deprived	Heisinger *et al.* (1973)
Onchomys torridus	desert	3180	4250	1710	—	2450	NaCl drink	Schmidt-Nielsen and Haines (1964)
Ondatra zibethica osyoosensis	semi-aquatic	1063	—	842	—	—	deprived	Zahn (1968)
Neofiber alleni	semi-aquatic	658	—	275	—	—	deprived	Pfeiffer (1970)
Microtus p. pennsylvanicus	mesic	1663	—	—	—	—	deprived	Heisinger *et al.* (1973)
M. o. ochrogaster	mesic	2544	—	—	—	—	NaCl inject.	Heisinger *et al.* (1973)
M. o. haydeni	mesic	2362	—	—	—	—	deprived	Heisinger *et al.* (1973)
Meriones shawii shawii	desert	2000	5000	—	—	—	anaesth.[b]	Rouffignac *et al.* (1969)
Psammomys obesus	desert	—	4950	—	—	1310	deprived	Schmidt-Nielsen and O'Dell (1961)
P. obesus	desert	—	6340	—	—	2850	—	Schmidt-Nielsen (1964)
P. obesus	desert	—	—	—	—	2720	deprived	Schmidt-Nielsen *et al.* (1961)
P. obesus	desert	1925	2765	—	—	—	anaesth.[b]	Morel *et al.* (1969)
MURIDAE								
Mus musculus feral	mesic	4720	7000	2910	—	3990	restricted	Haines *et al.* (1973a)
M. musculus feral	mesic	5438	—	3396	—	—	restricted	Haines and Schmidt-Nielsen (1967)

TABLE VI (continued)

Species	Habitat	Osmolality		Urea concentration		Water ration	Reference
		mean	max	mean	max		
M. musculus feral	salt-marsh	—	6200	—	—	deprived	Greene and Fertig (1972)
Leggadina hermannsburgensis	desert	4710	8970	2760	3920	deprived	MacMillen and Lee (1967); MacMillen et al. (1972)
Notomys alexis	desert	6550	9370	3430	5430	deprived	MacMillen and Lee (1967, 1969)
N. cervinus	desert	3720	4920	2500	3140	deprived	MacMillen and Lee (1967, 1969)
Acomys russatus	desert	—	—	4465	4800	deprived	Shkolnik and Borut (1969)
A. cahirinus	desert	—	—	4330	4700	deprived	Shkolnik and Borut (1969)
DIPODIDAE							
Jaculus jaculus	desert	—	6500	—	4320	—	Schmidt-Nielsen (1964)
HYSTRICOMORPHA ERETHIZONTIDAE							
Erethizon dorsatum	moist	1195	—	269	—	deprived	Pfeiffer (1970)
CHINCHILLIDAE							
Chinchilla laniger	arid	3505	7599	—	—	deprived	Weisser *et al.* (1970)
C. laniger	arid	2006	—	—	—	deprived	Gutman and Beyth (1970)
C. laniger	arid	—	6000	—	—	deprived	Prong *et al.* (1969)
CAPROMYIDAE							
Myocastor coypu	semi-aquatic	741	—	499	—	deprived	Pfeiffer (1970)

[a] Calculated from the data of Soholt (1975) and urine specific gravity from Silverstein (1961).
[b] Anaesthetized, non-diuretic animals injected with 0·9% NaCl.

urine to plasma osmotic ratio of 24·6 may be reached (MacMillen and Lee, 1967, 1969) and in *Leggadina hermannsburgensis* the slightly lower urine osmolality of 8970 mosmol but an even higher urine : plasma osmotic ratio of 26·8 may be reached (MacMillen and Lee, 1967; MacMillen *et al.*, 1972). The elimination of excretory products in a highly concentrated urine will reduce the urinary water loss thus being to the animal's advantage under the conditions of water scarcity which generally prevail in deserts.

In comparison, rodents from moist habitats seem to have a greatly reduced capacity for urinary water conservation. Only too few studies have been concerned with these rodents. They seem however, to be unable to elaborate a urine more concentrated than about 1000 mosmol thus having a high obligatory water loss. Just from the poor concentrating power of their kidneys rodents from most habitats seem to be excluded from extending their range into more arid habitats. Daily rhythms in urine osmolality and urine electrolyte concentrations were found in the hystricomorph rodents *Chinchilla laniger*, *Lagostomus maximus*, *Octodon degus*, *Galea musteloides*, *Cavia apera* and *C. porcellus* being confined to metabolic cages (Bellamy and Weir, 1972).

Sperber (1944) pointed out that desert rodents have an unusually high relative thickness of the kidney medulla and suggested this to be causative to the formation of concentrated urine in these forms. This is consonant with the current theory of urine concentration by the action of a countercurrent multiplier along the loop of Henle. A longer multiplier system would build up a higher interstitial osmolality in the papilla and thus allow urine of equally high osmolality to drain into the renal pelvis. Desert rodents producing extremely concentrated urine also have exceptionally long renal papillae extending well into the ureter (e.g. *Chinchilla laniger*, Weisser *et al.*, 1970; *Dipodomys merriami*, Schmidt-Nielsen, 1964; Soholt, 1975; *Leggadina hermannsburgensis*, MacMillen *et al.*, 1972; *Notomys alexis* and *N. cervinus*, MacMillen and Lee, 1969; *Psammomys obesus*, Abdallah and Tawfik, 1971; Rouffignac *et al.*, 1973).

Low urine osmolality of rodents from moist habitats (Table VI) is correlated with a low relative thickness of the renal medulla and a lack of the inner medullary zone (e.g. *Aplodontia rufa*, Schmidt-Nielsen and Pfeiffer, 1970; *Castor canadensis*, Schmidt-Nielsen and O'Dell, 1961). Among cricetid rodents (*Microtus*, *Peromyscus* and *Reithrodontomys*) there appears to be a direct relationship between the percentage medullary thickness and the ability to concentrate urine, at least congenerically (Heisinger *et al.*, 1973). The anatomical and physiological data correlate well with the judged aridity of the microhabitat of these cricetid rodents and it seems as if small changes in water availability might provide the selection pressure for renal structural and functional changes. From a survey of 15 rodents species sampled in the field, Schmid (1972) found a close correlation between the urine osmolality and the relative medullary area of the kidneys. He proposed that the increased capacity for concentrating urine in small, nocturnal rodents may have evolved as a mechanism to compensate for increased rates of pulmocutaneous water loss during the nights.

c. *Urine urea.* The urea concentration is much higher in urine of desert rodents than in urine of rodents from moist habitats (Table VI). This minimizes the urinary water loss necessary for urea excretion in desert rodents and may be regarded as an adaptation to the water scarcity. As shown by several studies (Schmidt-Nielsen and Kerr, 1970) urea has a pronounced effect on the concentrating capacity of rodent kidneys as well as of other mammalian kidneys. Upon water restriction there is a parallel increase in urine urea concentration and urine osmolality in feral *M. musculus* (Haines et al., 1973a). When drinking NaCl solutions of increasing molarities the canyon mouse, *Peromyscus crinitus stephensi*, shows the same correlation (Fig. 4). Upon water restriction and upon sea water drinking the salt-marsh harvest mouse, *Reithrodontomys raviventris halicoetes*, increases urine osmolality and urine urea concentration by 277% and 266%, and by 240% and 158% respectively (Haines, 1964). When the protein content of the food is increased several rodent species show an increase in urine osmolality (e.g. *Meriones shawi*, Grisard-Oberschall, 1968; feral *M. musculus*, Haines and Schmidt-Nielsen, 1967; *Onychomys torridus*, Schmidt-Nielsen and Haines, 1964). Increased urine osmolality has also been induced by urea administration in white rats, *R. norvegicus*, which have been deprived of dietary protein (Pennell et al., 1975; Heuer et al., 1974). Urea seems to enhance urinary concentration by accumulating in the interstitial fluid of the inner medulla causing an enlarged osmotic water removal from the nephridial fluid. It is unsettled whether the water abstraction is from the collecting ducts (Heuer et al., 1974) or from the descending limb of the loop of Henle (Pennell et al., 1975).

In some rodents (*Aplodontia rufa*, Schmidt-Nielsen and Pfeiffer, 1970; *Myocastor coypu* and *Neofiber alleni*, Pfeiffer, 1970; *Ondatra zibethica osoyoosensis*, Zahn, 1968; *Castor canadensis* and *Psammomys obesus*, Schmidt-Nielsen et al., 1961) increased dietary protein intake or dietary urea administration does not increase urine osmolality. In these animals there is no positive correlation between urine osmolality and urine urea concentration. Rather, in *Aplodontia rufa* (Schmidt-Nielsen and Pfeiffer, 1970) and in *Castor canadensis* and *Psammomys obesus* (Schmidt-Nielsen et al., 1961) it seems as if there is a fixed osmotic ceiling of the urine and that urine urea and urine electrolyte concentrations vary inversely

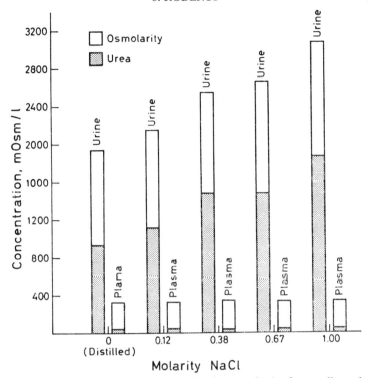

Fig. 4. Mean osmolality and urea concentration in plasma and urine from acclimated canyon mouse, *Peromyscus crinitus stephensi*, while drinking distilled water and increased concentrations of NaCl (from Abbott, 1971).

when the urine is maximally concentrated. It has been suggested (Pfeiffer, 1970) that the ability of urea to increase the concentrating capacity of the kidney depends upon the presence of certain pelvic and medullary structures.

d. *Antidiuretic hormone.* Although the mechanism for water conservation exists in the kidney itself, it is the function of the antidiuretic hormone (ADH) to activate it. In rodents, the typical ADH seems to be arginine-vasopressin (Bentley, 1971). When subjected to dehydrating conditions ADH is secreted from its storage site in the neural lobe of the pituitary gland and increasing amounts appear in the blood. In the kidney the ADH exerts its effect by increasing the water permeability along the full extent of the collecting system of the nephron (Grantham, 1974). This promotes water withdrawal from the collecting duct fluid thus concentrating and reducing the volume of the final urine. It has been shown that desert rodents generally have a larger capacity for ADH synthesis and higher plasma levels of ADH than rodents from mesic and moist habitats (Bentley, 1971).

Upon water deprivation, albino *M. musculus* and the desert rodents, *Acomys russatus* and *A. cahirinus* show a temporal depletion of neurosecretory material of the neural lobe of the pituitary gland followed by repletion after about two weeks of water deprivation (Castel and Abraham, 1969). Concomitantly, dehydrated animals showed a progressive increase in the number of actively synthetizing neuronal cell bodies of the supraoptic nucleus in the hypothalamus. Hatton (1972) showed that xeric rodents have a higher proportion of cells with multiple nucleoli in the supraoptic nucleus than do rodents from mesic habitats, suggesting that the number of nucleoli is related to the production of ADH. Such data are indicative of an acclimation to water shortage by increased ADH synthesis and increased plasma levels of ADH thus increasing the water conserving capacity of the kidney. Possibly the increase in circulating ADH has a further water conserving effect by reducing the transepidermal water loss (Hattingh, 1975). A seasonal variation in the plasma level of antidiuretic substances was found by El-Husseini and Haggag (1974) in the desert rodents *Jaculus jaculus* and *Gerbillus gerbillus*. Maximal antidiuretic activity of the plasma was found in animals during the summer when the advantage of an effective water economy would be greatest. When water-stressed in the laboratory on a diet of oven-dried barley and no water both summer and winter animals showed a further increase in the antidiuretic activity of the plasma. The same two species have also been shown to exhibit maximal neurosecretory activity of the hypothalamic neurohypophyseal system during the summer season suggesting maximal turnover of ADH at this time (Khalil and Taufic, 1964).

D. FAECAL WATER LOSS

Faecal water loss makes up a small percentage of the total water loss in rodents (Schmidt-Nielsen and Schmidt-Nielsen, 1951; Haines *et al.*, 1973a; Soholt, 1975). Under conditions of water shortage, however, any water loss will be detrimental and a reduction of the faecal water loss would be of advantage to the animal. Faecal water content and faeces egestion of various rodents are given in Table VII. Although the faecal water content varies between species, regardless of habitat, there is invariably a decrease in the water content upon water deprivation of the animal. Moreover, the observed decrease in the daily amount of faeces egested will magnify the reduction in faecal water loss. The data point to a control of faecal water loss in rodents. However, little physiological evidence is available characterizing the mechanisms involved.

The kangaroo rat, *Dipodomys merriami*, has habitual coprophagy and this, as was pointed out by Schmidt-Nielsen (1963, 1964), would work towards a greater economy in the water expenditure of the animal. Coprophagy is known from a number of rodents but seems to have been of little concern in studies of the osmotic regulation of these animals. It would be of interest to know whether coprophagy is related to body dehydration in rodents. In the rabbit which has obligatory coprophagy, Staaland (1974) found a marked reduction in the content of water, Na and K of the faecal matter during its second passage through the colon. Acutely water-deprived feral *M. musculus* showed a significant decrease in faeces production but had an equal food consumption when compared to animals given water *ad libitum* (Haines *et al.*, 1973a). This could reflect increased coprophagous activity of the water-deprived *M. musculus*.

Attention has been paid to the post-caecal spiral in the reabsorption of water and ions from the digestive tract of cricetid rodents (Lange and Staaland, 1970). The post-caecal spiral is located opposite to the caecum at the junction between ileum and colon, and consists of a descending and an ascending part. This structure was found in eight herbivorous cricetid rodents but not in the murids studied (*M. musculus*, *R. norvegicus* and *Apodemus sylvaticus*). The post-caecal spiral was most highly developed in those animals naturally feeding on a diet poor in mineral ions: i.e. *Lemmus lemmus*, *Myopus schisticolor* and *Dicrostonyx torquatus* feeding preferentially on mosses. Lange and Staaland (1970) and Staaland (1975) found for the Norwegian lemming, *L. lemmus*, that Na reabsorption from the intestinal content seemed especially associated with the post-caecal spiral, while K and water reabsorption mainly occurred from the proximal colon. In the white rat, water and ions were chiefly reabsorbed in the proximal colon. The amount of water reabsorbed from the caecum-colon region of the lemming and the rat could apparently be accounted for by the combined reabsorption of Na and K thus suggesting a solute linked transport for the water reabsorption.

Macromolecules such as proteins were suggested to act as osmotic effector substances in the proximal colon of the white rat (Lange and Staaland, 1971). Katz (1973) found the water content of the whole gastrointestinal tract with its content to be reduced upon water deprivation in various rodents at 37°C. The effect was more pronounced for xerophilic than for mesophilic rodents.

E. LACTATION

Rodents, especially myomorphs, are reknown for their reproductive capacity having frequent births and large litters (Walker, 1964). A

TABLE VII

Faecal water content and faeces egestion in various rodents given dry food *ad libitum* and either deprived of water (D) or given water *ad libitum* (A)

Species	Habitat	Body wt (g)	Faecal water %	Faeces mg dry wt g^{-1} day^{-1}	Water ration	Food	Reference
SCIUROMORPHA							
HETEROMYIDAE							
Dipodomys merriami	desert	—	45	—	D	pearled barley	Schmidt-Nielsen and Schmidt-Nielsen (1951)
D. merriami	desert	34	40	4·1	D	rolled oats	Soholt (1975)
MYOMORPHA							
CRICETIDAE							
Peromyscus leucopus	mesic	23	44	30	A	rolled oats	Getz (1968)
P. truei	arid	21	57	—	A	mouse chow	Bradford (1974)
P. truei	arid	12	52	—	D^a	mouse chow	Bradford (1974)
Mesocricetus auratus	arid	—	74	—	A	wheat seeds	Katz (1973)
M. auratus	arid	—	37	—	D	wheat seeds	Katz (1973)
Lemmus lemmus	moist	—	51	—	A	mixed diet	Lange and Staaland (1970)
Clethrionomys gapperi	moist	23	49	24	A	rolled oats	Getz (1968)
Gerbillus dasyurus	desert	21	64	—	A	wheat seeds	Katz (1973)
G. dasyurus	desert	21	38	—	D	wheat seeds	Katz (1973)
Meriones crassus	desert	52	52	—	A	wheat seeds	Katz (1973)
M. crassus	desert	90	21	—	D	wheat seeds	Katz (1973)

MURIDAE							
Rattus norvegicus	moist	—	68	—	D	pearled barley	Schmidt-Nielsen and Schmidt-Nielsen (1951)
R. norvegicus	moist	182	70	—	A	purina chow	Katz (1973)
R. norvegicus	moist	137	52	—	D	purina chow	Katz (1973)
R. norvegicus	moist	—	67	—	A	mixed diet	Lange and Staaland (1970)
Mus musculus	mesic	21	61	8·2	A	dry chow	Haines *et al.* (1973a)
M. musculus	mesic	—	39	5·2	D	dry chow	Haines *et al.* (1973a)
M. musculus	mesic	—	38	—	D	dry chow	Fertig and Edmonds (1969)
Leggadina hermannsburgensis	desert	12	50	—	D	bird seeds	MacMillen and Lee (1967)
Notomys alexis	desert	32	—	8·4	A	bird seeds	MacMillen and Lee (1969)
N. alexis	desert	32	49	2·8	D	bird seeds	MacMillen and Lee (1969)
N. cervinus	desert	34	—	5·3	A	bird seeds	MacMillen and Lee (1969)
N. cervinus	desert	34	52	3·8	D	bird seeds	MacMillen and Lee (1969)
DIPODIDAE							
Jaculus jaculus	desert	69	44	—	A	wheat seeds	Katz (1973)
J. jaculus	desert	58	36	—	D	wheat seeds	Katz (1973)
J. orientalis	desert	152	47	5·3	A	wheat seeds	Kirmiz (1962)
J. orientalis	desert	147	42	3·5	D	wheat seeds	Kirmiz (1962)

[a] Minimum water for continued survival.

female *M. musculus* nursing a litter of nine to 12 pups will be supplying milk to a combined mass of young three to four times greater than her own weight by the end of lactation (Smith and McManus, 1975). The mean milk yield of *R. norvegicus* with 12 pups is 42 ml day^{-1} on the tenth day of lactation (Hanwell and Linzell, 1972). This large drainage of water and ions associated with lactation would be expected to influence markedly on the osmotic regulation of the female rodent. Few investigations, however, have been concerned with this problem

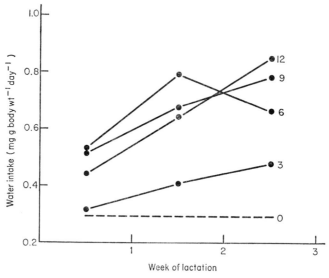

Fig. 5. Water intake of mice, *Mus musculus*, nursing litters from three to 12 pups. The dashed line represents the intake of non-reproductive, non-lactating controls (from Smith and McManus, 1975).

although it was explicitly presented by Schmidt-Nielsen (1964) in his monograph on desert animals.

The water intake of lactating rodents is greatly increased relative to non-reproductive females: *Peromyscus maniculatus rufinus* with 5 pups has an increase of 375% during the 10 last days of lactation and *P. t. truei* with three pups an increase of 200% during 4 weeks of lactation (Douglas, 1969). *Mus musculus* with a litter of 12 has an increase of almost 300% in the third week of lactation (Fig. 5). The high rate of water intake of females of the ground squirrel, *Citellus tridecemlineatus*, in June (maximum values 0·72 ml g^{-1} day^{-1}) as compared to September (maximum values 0·09 ml g^{-1} day^{-1}) found by Armitage and Shulenberger (1972) is probably caused by the fact that seven of the 13 females tested had lactating litters which weaned in the end of June to early July. In the ground squirrel *Spermophilus tereticaudatus*, the breeding season is correlated with the rainy season and the litter size correlated with rainfall size (Reynolds and Turkowski, 1972). For the kangaroo rat, *Dipodomys merrriami*, the availability of green vegetation is determining for the reproductive period (Bradley and Mauer, 1971), and for the rats, *R. rattus* and *R. exulans*, the reproductive activity is controlled by environmental factors (Tamarin and Malecha, 1972). These data clearly indicate that reproductive patterns have been selected for tending to reduce the osmotic burden of lactation. However, it should be interesting to know how renal and extrarenal mechanisms are involved in the regulation of body water and ions during lactation.

Recently, Baverstock and Green (1975) found a high degree of water recycling between the lactating rodent and the suckling young. By consuming the urine and faeces of the young, the lactating females of *Notomys alexis*, *N. cervinus*, *N. mitchellii*, *Pseudomys australis*, *Mus musculus* and *Rattus norvegicus* were found to regain about one-third of the water lost in the milk. Due to the larger concentrating power of the kidneys of the mother relative to that of the young (e.g. the urine osmolality of the young of *N. alexis* is only about 20% of that of the mother, Baverstock and Green, 1975) this behaviour results in a net gain of water for the mother. Thus, for desert rodents under conditions of water scarcity such recycling of water will greatly reduce the problem of osmotic regulation during lactation. From a study of the milk of these same rodent species Baverstock *et al.* (1976) concluded that the milk concentration was no important factor on the water balance of the lactating female. The water content (free water plus metabolic water) of the milk was unchanged between animals kept on *ad libitum* or restricted water intakes.

IV. CONCLUSIONS

Rodents are the most successful group of mammals in terms of number of species and number of specimens. Also, their ubiquitousness in terrestrial habitats varying greatly in water availability attests to the adaptability of their osmotic regulation. It emerges from the works reviewed in this chapter that several mechanisms co-operate to maintain the osmotic homeostasis of the body fluids of a rodent. It seems that these mechanisms are qualitatively similar in rodents from moist and from xeric habitats, their distinction being in quantitative terms.

The correlation between environmental aridity and water requirement or water turnover rate of rodents (Table II) indicates that the water availability acts as a selection pressure towards a greater water economy of the animal. Some rodents succeed in maintaining the water balance on a diet of dry seeds, most rodents, however, are dependent upon an additional source of water. Through the preference for succulent food a rodent may evade the apparent aridity of its habitat. Increased knowledge of the micro-environment, behaviour and food preference of each rodent species is necessary to characterize the actual water availability to the animal in its niche. Seasonal changes in food type composition may be regarded as behavioural adaptations to optimize the water balance of the animal.

Associated with the terrestrial life rodents experience an obligatory evaporative water loss. Depending upon the magnitude of the evaporative water loss an osmotic stress of variable size is transferred to the water-conserving mechanisms of the kidneys. In rodents, the evaporative water loss is lower in xeric forms than in species from more humid habitats (Table V). The evaporative water loss seems to be physiologically controlled although little is known about the mechanisms governing this regulation. Upon prolonged water restriction the evaporative water loss may be reduced, possibly by a decrease in the water loss through cutaneous routes. The pulmonary water loss seems to be lower in rodents than in other mammals, a feature which has been ascribed to the presence of a countercurrent heat and water exchanger in the nasal passageways of rodents.

The kidneys are the main osmoregulating organs in rodents, adjusting both water and solute excretion over wide ranges. Although being primitively simple in morphology the rodent kidney has developed into the most powerful water-conserving organ found among mammals. High concentrating power is associated with a long renal papilla and a high degree of regularity in the microanatomical arrangement of nephridial and vascular structures. The structural arrangement is favourable to effective countercurrent exchange functions. The concentrating power of rodent kidneys is correlated with the water availability of the microhabitat. Thus, rodents inhabiting xeric regions can produce highly concentrated urine while rodents from moist habitats are able to elaborate only moderately concentrated urine. The elimination of excretory products in a highly concentrated urine will reduce the urinary water loss thus being to the advantage of a rodent from a xeric habitat. The high urine concentrations of rodents from xeric regions may partly be explained by chronically high plasma levels of ADH and a high capacity for ADH synthesis. Increased ADH synthesis seems to be part of the acclimation process towards long-term water shortage.

The faecal water loss in rodents is small and generally plays a minor role in the osmotic balance of the animal. Under conditions of water shortage, however, any water loss would be detrimental and a reduction in the faecal water loss would be to the animal's advantage. The few available data point to a reduced faecal water loss in water-deprived rodents and the phenomenon deserves further study.

REFERENCES

Abbott, K. D. (1971). Water economy of the canyon mouse *Peromyscus crinitus stephensi*. *Comp Biochem. Physiol.* **38A**, 37–52.
Abdallah, A. and Tawfik, J. (1969). The anatomy and histology of the kidney of sand rats (*Psammomys obesus*). *Z. Versuchtierk* **11**, 261–275.
Abdallah, A. and Tawfik, J. (1971). Effects of sodium chloride on the water consumption of sand rats (*Psammomys obesus*). *Z. Versuchtierk.* **13**, 150–157.
Adolph, E. F. (1943). "Physiological Regulations." Cattell, Lancaster.
Adolph, E. F. (1949). Quantitative relations in the physiological constituents of mammals. *Science N.Y.* **109**, 579–585.
Anderson, S. H. (1970). Effect of temperature on water loss and CO_2 production of *Perognathus parvus*. *J. Mammal* **51**, 619–620.
Andrews, P. M. and Porter, K. R. (1974). A scanning electron microscope study of the nephron. *Am. J. Anat.* **140**, 81–116.
Arakawa, M. (1970). A scanning electron microscope study of the glomerulus of normal and nephrotic rats. *Lab. Invest.* **23**, 489–496.
Armitage, K. B. and Shulenberger, E. (1972). Evidence for a circannual metabolic cycle in *Citellus tridecemlineatus*, a hibernator. *Comp. Biochem. Physiol.* **42A**, 667–688.
Babero, B. B., Yousef, M. K., Wawerna, J. C. and Bradley, W. G. (1973). Comparative histology of the respiratory apparatus of three desert rodents and the albino rat. A view on morphological adaptations. *Comp. Biochem. Physiol.* **44A**, 585–597.
Batsli, G. and Pitelka, F. (1971). Condition and diet of cycling populations of the California vole, *Microtus californicus*. *J. Mammal.* **52**, 141–163.
Baudinette, R. V. (1972a). Energy metabolism and evaporative water loss in the California ground squirrel. Effects of burrow temperature and water vapour pressure. *J. comp. Physiol.* **81**, 57–72.
Baudinette, R. V. (1972b). The impact of social aggregation on the respiratory physiology of Australian hopping mice. *Comp. Biochem. Physiol.* **41A**, 35–38.
Baverstock, P. and Green, B. (1975). Water recycling in lactation. *Science N.Y.* **187**, 657–658.
Baverstock, P., Spencer, L. and Pollard, C. (1976). Water balance of small lactating rodents—II. Concentration and composition of milk of females on *ad libitum* and restricted water intakes. *Comp. Biochem. Physiol.* **53A**, 47–52.
Bellamy, D. and Weir, B. J. (1972). Urine composition of some hystericomorph rodents confined to metabolic cages. *Comp. Biochem. Physiol.* **42A**, 759–771.
Bentley, P. J. (1971). "Endocrines and Osmoregulation." Springer Verlag, Berlin, Heidelberg and New York.
Birney, E. C. and Twomey, S. L. (1970). Effect of sodium chloride on water consumption, weight, and survival in the woodrats, *Neotoma micropus* and *Neotoma floridana*. *J. Mammal* **51**, 372–375.
Boice, R. (1972). Water addiction in captive desert rodents. *J. Mammal* **53**, 395–398.
Boice, R. and Witter, J. A. (1970). Water deprivation and activity in *Dipodomys ordii* and *Meriones unguiculatus*. *J. Mammal.* **51**, 615–618.

Bradford, D. F. (1974). Water stress of free-living *Peromyscus truei*. *Ecology* **55**, 1407–1414.

Bradley, W. G. and Mauer, R. A. (1971). Reproduction and food habits of Merriam's kangaroo rat, *Dipodomys merriami*. *J. Mammal.* **52**, 497–507.

Bradley, W. G., Yousef, M. K. and Scott, I. M. (1975). Physiological studies of the rock pocket mouse, *Perognathus intermedius*. *Comp. Biochem. Physiol.* **50A**, 331–337.

Breyen, L. J., Bradley, W. G. and Yousef, M.K. (1973). Physiological and ecological studies on the chiseltoothed kangaroo rat, *Dipodomys microps*. *Comp. Biochem. Physiol.* **44A**, 543–555.

Brower, J. E. and Cade, T. J. (1966). Ecology and physiology of *Napaeozapus insignis* (Miller) and other woodland mice. *Ecology* **47**, 46–53.

Bulger, R. A., Siegel, F. L. and Pendergrass, R. (1974). Scanning and transmission electron microscopy of the rat kidney. *Am. J. Anat.* **139**, 483–518.

Burg, M. B. and Green, N. (1973). Function of the thick ascending limb of Henle's loop. *Am. J. Physiol.* **224**, 659–668.

Carleton, M. C. (1973). A survey of gross stomach morphology in New World Cricetinae (Rodentia, Muroidea), with comments on functional interpretations. *Misc. Publs Mus. Zool. Univ. Mich.* No. 146, 43 pp.

Carpenter, R. E. (1966). A comparison of thermoregulation and water metabolism in the kangaroo rats *Dipodomys agilis* and *Dipodomys merriami*. *Univ. Calif. Publ. Zool.* **78**, 1–36.

Castel, M. and Abraham, M. (1969). Effects of a dry diet on the hypothalamic neurohypophyseal neurosecretory system in spiny mice as compared to the albino rat and mouse. *Gen. comp. Endocr.* **12**, 231–241.

Chew, R. M. (1955). The skin and respiratory water losses of *Peromyscus maniculatus sonoriensis*. *Ecology* **36**, 463–467.

Chew, R. M. (1965). Water metabolism of mammals. *In* "Physiological Mammalogy" (R. V. Mayer and R. G. Van Gelder, eds) Vol. II. Academic Press, New York and London, pp. 43–178.

Chew, R. M. and Dammann, A. E. (1961). Evaporative water loss of small vertebrates, as measured with an infrared analyzer. *Science N.Y.* **133**, 384–385.

Church, R. L. (1966). Water exchanges of the California vole, *Microtus californicus*. *Physiol. Zoöl.* **39**, 326–340.

Clapp, J. R. (1966). Renal tubular reabsorption of urea in normal and protein-depleted rats. *Am. J. Physiol.* **210**, 1304–1308.

Clapp, J. R. (1970). The effect of protein depletion on urea reabsorption by the kidney. *In* "Urea and the Kidney" (B. Schmidt-Nielsen and D. W. S. Kerr, eds). Excerpta Medica Foundation, Amsterdam, pp. 200–205.

Collins, B. G. (1973a). Physiological responses to temperature stress by an Australian murid, *Rattus lutreolus*. *J. Mammal.* **54**, 356–368.

Collins, B. G. (1973b). The ecological significance of thermoregulatory responses to heat stress shown by two populations of an Australian murid, *Rattus fuscipes*. *Comp. Biochem. Physiol.* **44A**, 1129–1140.

Collins, B. G. and Bradshaw, S. D. (1973). Studies on the metabolism, thermoregulation and evaporative water losses of two species of Australian rats, *Rattus villosissimus* and *Rattus rattus*. *Physiol. Zoöl.* **46**, 1–21.

Collins, J. C., Pilkington, T. C. and Schmidt-Nielsen, K. (1971). A model of respiratory heat transfer in a small mammal. *Biophys. J.* **11**, 886–914.

Dicker, S. E. (1970). "Mechanisms of Urine Concentration and Dilution in Mammals." Edward Arnold, London.

Dicker, S. E. and Eggleton, M. G. (1964). Renal function in the primitive mammal *Apoldontia rufa*, with some observations on squirrels. *J. Physiol. (Lond.)* **170**, 186–194.

Dieterich, H. J., Barrett, J. M., Kriz, W. and Bülhoff, J. P. (1975). The ultrastructure of the thin loop limbs of the mouse kidney. *Anat. Embryol.* **147**, 1–18.

Douglas, C. L. (1969). Comparative ecology of pinyon mice and deer mice in Mesa Verde National Park, Colorado. *Univ. Kans. Publs Mus. nat. Hist.* **18**, 421–504.

El-Husseini, M. and Haggag, G. (1974). Antidiuretic hormone and water conservation in desert rodents. *Comp. Biochem. Physiol.* **47A**, 347–350.

Ellerman, J. R. (1940). "The Families and Genera of Living Rodents" Vol. I. British Museum (nat. Hist.), London.

Ellerman, J. R. (1941). "The Families and Genera of Living Rodents" Vol. II. British Museum (nat. Hist.), London.

Ericsson, J. L. E. and Trump, B. F. (1969). Electron microscopy of the uriniferous tubules. *In* "The Kidney—Morphology, Biochemistry, Physiology" (C. Rouiller and A. F. Muller, eds) Vol. I. Academic Press, New York and London, pp. 351–447.

Ernst, C. H. (1968). Kidney efficiencies of three Pennsylvania mice. *Trans. Ky Acad. Sci.* **29**, 21–24.

Ewing, W. G. and Studier, E. H. (1973). A method for control of water vapour pressure and its effect on metabolism and body temperature in the *Mus musculus Comp. Biochem. Physiol.* **45A**, 121–125.

Fertig, D. S. and Edmonds, V. W. (1969). The physiology of the house mouse. *Sci. Am.* **221**, 103–110.

Fertig, D. S. and Layne, J. N. (1963). Water relationships in the Florida mouse. *J. Mammal.* **44**, 322–334.

Fisler, G. F. (1963). Effects of salt water on food and water consumption and weight of harvest mice. *Ecology* **44**, 604–608.

Forbes, R. B. (1967). Some aspects of the water economics of two species of chipmunks. *J. Mammal.* **48**, 466–468.

Frenkel, G., Shaham, Y. and Kraiger, P. F. (1972). Establishment of conditions for colony-breeding of the sand-rat *Psammomys obesus*. *Lab. Anim. Sci.* **22**, 40–47.

Getz, L. L. (1963). A comparison of the water balance of the prairie and meadow voles. *Ecology* **44**, 202–207.

Getz, L. L. (1968). Influence of water balance and microclimate on the local distribution of the redback vole and white-footed mouse. *Ecology* **49**, 276–286.

Glen, M. E. (1970). Water relations in three species of deer mice (*Peromyscus*). *Comp. Biochem. Physiol.* **33**, 231–248.

Grantham, J. J. (1974). Action of antidiuretic hormone in the mammalian kidney. *In* "Kidney and Urinary Tract Physiology" (K. Thurau, ed.) MTP Int. Rev. Sci., Physiology. Series one, Vol. VI. Butterworths, London, pp. 247–272.

Greene, J. R. and Fertig, D. S. (1972). Water sources for house mice living in salt marshes. *Physiol. Zoöl.* **45**, 125–129.

Grisard-Operschall, P. (1968). Untersuchungen über die Nierenfunktionen von *Meriones shawii*. *Revue suisse Zoöl.* **75**, 1–41.

Gutman, Y. and Beyth, Y. (1970). *Chinchilla laniger*—discrepancy between concentrating ability and kidney structure. *Life Sci.* **9**, 37–42.

Haines, H. (1964). Salt tolerance and water requirements in the salt-marsh harvest mouse. *Physiol. Zoöl.* **37**, 266–272.

Haines, H., Ciskowski, C. and Harms, V. (1973a). Acclimation to chronic water restriction in the wild house mouse *Mus musculus*, *Physiol. Zoöl.* **46**, 110–128.

Haines, H., Howard, B. and Setchell, C. (1973b). Water content and distribution of tritiated water in tissues of Australian desert rodents. *Comp. Biochem. Physiol.* **45A**, 787–792.

Haines, H. and Schmidt-Nielsen, K. (1967). Water deprivation in wild house mice. *Physiol. Zoöl.* **40**, 424–431.

Haines, H. and Shield, C. F. (1971). Reduced evaporation in house mice (*Mus musculus*) acclimated to water restriction. *Comp. Biochem. Physiol.* **39A**, 53–61.

Haines, H., Shield, C. F. and Twitchell, C. (1969). Reduced evaporation from rodents during prolonged water restriction. *Life Sci.* **8**, 1063–1068.

Hanwell, A. and Linzell, J. L. (1972). A simple technique for measuring the rate of milk secretion in the rat. *Comp. Biochem. Physiol.* **43A**, 259–270.

Hattingh, J. (1972a). A comparative study of transepidermal water loss through the skin of various animals. *Comp. Biochem. Physiol.* **43A**, 715–718.

Hattingh, J. (1972b). The correlation between transepidermal water loss and the thickness of epidermal components. *Comp. Biochem. Physiol.* **43A**, 719–722.

Hattingh, J. (1975). The influence of hormones and blood flow on transepidermal water loss. *Comp. Biochem. Physiol.* **50A**, 439–442.

Hatton, G. J. (1972). Supraoptic nuclei of rodents adapted for mesic and xeric environments: Number of cells, multiple nucleoli, and their distribution. *J. comp. Neurol.* **145**, 43–60.

Heisinger, J. F., King, T. S., Halling, H. W. and Fields, B. L. (1973). Renal adaptations to macro- and microhabitats in the family Cricetidae. *Comp. Biochem. Physiol.* **44A**, 767–774.

Heuer, L. J., Wester, H. and Voss, P. (1974). Die Wirkung von Harnstoff auf die Wasserresorption der saltskonsentrierenden Niere unter Gleichgewichtsbedingungen. *Res. exp. Med.* **162**, 333–339.

Holleman, D. F. and Dieterich, R. A. (1973). Body water content and turnover in several species of rodents as evaluated by the tritiated water method. *J. Mammal.* **54**, 456–465.

House, E. W., Pfeiffer, E. W. and Braun, H. A. (1963). Influence of diet on urine concentration in *Aplodontia rufa* and the rabbit. *Nature* **199**, 181–182.

Hudson, J. W. (1962). The role of water in the biology of the antelope ground squirrel *Citellus leucurus*. *Univ. Calif. Publs Zool.* **64**, 1–56.

Hudson, J. W. and Deavers, D. R. (1973a). Metabolism, pulmocutaneous water loss and respiration of eight species of ground squirrels from different environments. *Comp. Biochem. Physiol.* **45A**, 69–100.

Hudson, J. W. and Deavers, D. R. (1973b). Thermoregulation at high ambient temperatures of six species of ground squirrels (*Spermophilus* ssp) from different habitats. *Physiol. Zoöl.* **46**, 95–109.

Hudson, J. W. and Rummel, J. A. (1966). Water metabolism and temperature regulation of the primitive heteromyids, *Liomys salvani* and *Liomys irroratus*. *Ecology* **47**, 345–354.

Jackson, D. C. and Schmidt-Nielsen, K. (1964). Countercurrent heat exchange in the respiratory passages. *Proc. nat. Acad. Sci.* **51**, 1192–1197.

Jameson, E. W. (1952). Food of the deer mice, *Peromyscus maniculatus* and *Peromyscus boylei*, in the northern Sierra Nevada, California. *J. Mammal.* **33**, 50–60.

Jamison, R. L. (1974). Countercurrent systems. In "Kidney and Urinary Tract Physiology" (K. Thurau, ed.) MTP Int. Rev. Sci., Physiology, Series one, Vol. VI. Butterworths, London, pp. 199–245.

Jamison, R. L., Buerkert, J. and Lacy, F. (1973). A micropuncture study of Henle's thin loop in Brattleboro rats. *Am. J. Physiol.* **224**, 180–185.

Jenkinson, D. M. (1970). The distribution of nerves, monoamine oxidase and cholinesterase in the skin of the guinea-pig, hamster, mouse, rabbit and rat. *Res. vet. Sci.* **11**, 60–70.

Jenkinson, D. M. (1973). Comparative physiology of sweating. *Br. J. Derm.* **88**, 397–406.

Katz, U. (1973). The effects of water deprivation and hypertonic salt injection on several rodent species compared with the albino rat. *Comp. Biochem. Physiol.* **44A**, 473–485.

Khalil, F. and Taufic, J. (1964). Seasonal changes in the hypothalamo-hypophysial neurosecretory system of desert rodents. *J. Endocr.* **29**, 251–254.

Kirmiz, J. P. (1962). "Adaptation to Desert Environment." Butterworths, London.

Kokko, J. P. (1970). Sodium chloride and water transport in the descending limb of Henle. *J. clin. Invest.* **49**, 1838–1846.

Kokko, J. P. (1972). Urea transport in the proximal tubule and the descending limb of Henle. *J. clin. Invest.* **51**, 1099–2008.

Kokko, J. P. and Rector, F. C. (1972). Countercurrent multiplication system without active transport in inner medulla. *Kidney Int.* **2**, 214–223.

Kriz, W. (1967). Der architektonische und funktionelle Aufbau der Rattenniere. *Z. Zellforsch. Mikrosk. Anat.* **82**, 495–535.

Kriz, W. and Koepsell, H. (1974). The structural organization of the mouse kidney *Z. anat. Entwickl.-Gesch.* **144**, 137–163.

Kriz, W. and Lever, A. F. (1969). Renal countercurrent mechanisms, structure and function. *Am. Heart J.* **78**, 101–118.

Kriz, W., Schnermann, J. and Koepsell, H. (1972). The position of short and long loops of Henle in the rat kidney. *Z. anat. Entwickl.-Gesch.* **138**, 301–319.

Krizman, E. B. (1974). Ecological relationships of *Peromyscus maniculatus* and *Perognathus parvus* in eastern Washington. *J. Mammal.* **55**, 172–188.

Landry, S. O. (1970). The rodentia as omnivores. *Q. Rev. Biol.* **45**, 351–372.

Lange, R. and Staaland, H. (1970). Adaptations of the caecum-colon structure of rodents. *Comp. Biochem. Physiol.* **35**, 905–919.

Lange, R. and Staaland, H. (1971). On the mechanisms of water absorption in the rat colon. *Comp. Biochem. Physiol.* **40A**, 823–831.

Lasiewski, R. C., Acosta, A. L. and Bernstein, H. M. (1966). Evaporative water loss in birds—I. Characteristics of the open flow method of determination, and their relation to estimates of thermoregulatory ability. *Comp. Biochem. Physiol.* **19**, 445–457.

Latta, H., Maunsbach, A. B. and Osvaldo, L. (1967). The fine structure of renal tubules in cortex and medulla. *In* "Ultrastructure in Biological Systems" (A. J. Dalton and F. Haguenau, eds) Vol II. Academic Press, New York and London, pp. 1–56.

Lee, A. K. (1963). The adaptations to arid environments in wood rats of the genus *Neotoma*. *Univ. Calif. Publs Zool.* **64**, 57–96.

Leighly, J. (1937). A note on evaporation. *Ecology* **18**, 180–189.

McManus, J. J. (1972a). Water relations and food consumption of the Mongolian gerbil, *Meriones unguiculatus*. *Comp. Biochem. Physiol.* **43A**, 959–967.

McManus, J. J. (1972b). Water relations of the chinchilla *Chinchilla laniger*. *Comp. Biochem. Physiol.* **41A**, 445–450.

McManus, J. J. (1974). Bioenergetics and water requirements of the redback vole. *Clethrionomys gapperi*. *J. Mammal.* **55**, 30–44.

McNab, B. K. (1966). The metabolism of fossorial rodents: a study of convergence *Ecology* **47**, 712–733.

Macfarlane, W. V., Howard, B., Haines, H., Kennedy, P. J. and Sharpe, C. M. (1971). Hierarchy of water and energy turnover of desert mammals. *Nature* **234**, 483–484.

Machin, J. (1964). The evaporation of water from *Helix aspersa* II. Measurement of air flow and diffusion of water vapour. *J. exp. Biol.* **41**, 771–781.

Macmillen, R. E. (1964a). Population, ecology, water relations, and social behaviour of a southern California semidesert rodent fauna. *Univ. Calif. Publ. Zool.* **71**, 1–66.

Macmillen, R. E. (1964b). Water economy and salt balance in the western harvest mouse. *Reithrodontomys megalotis. Physiol. Zoöl.* **37**, 45–56.

Macmillen, R. E. and Lee, A. K. (1967). Australian desert mice: independence of exogenous water. *Science N.Y.* **158**, 383–385.

Macmillen, R. E. and Lee, A. K. (1969). Water metabolism of Australian hopping mice. *Comp. Biochem. Physiol.* **28**, 493–514.

Macmillen, R. E. and Lee, A. K. (1970). Energy metabolism and pulmocutaneous water loss of Australian hopping mice. *Comp. Biochem. Physiol.* **35**, 355–369.

Macmillen, R. E., Baudinette, R. V. and Lee, A. K. (1972). Water economy and energy metabolism of the sandy inland mouse, *Leggadina hermannsburgensis. J. Mammal.* **53**, 529–539.

Mansfield, K. I. (1970). The Norwegian lemming (*Lemmus lemmus* L.) in captivity. *J. Inst. Anim. Techns* **21**, 33–49.

Marsh, D. J. (1971). Osmotic concentration and dilution of the urine. In "The Kidney" (C. Rouller and A. F. Muller, eds) Vol. III. Academic Press, New York and London, pp. 71–126.

Maxson, K. A. and Morton, M. L. (1974). Water and salt regulation in the antelope ground squirrel (*Ammospermophilus leucurus*). *Comp. Biochem. Physiol.* **47A**, 117–128.

Morel, F. and Rouffignac, C. de (1973). Kidney. *A. Rev. Physiol.* **35**, 17–54.

Morel, R., Rouffignac, C. de, Mars, D., Guinnebault, M. and Lechene, C. (1969). Etude par microponction de l'élaboration de l'urine II. Chez le *Psammomys* non diurétique. *Nephron* **6**, 553–570.

Morris, D. (1965). "The Mammals." Hodder and Stoughton, London.

Moy, R. M., Lesser, R. W. and Pfeiffer, E. W. (1972). Urine concentrating ability of arousing and normothermic ground squirrels (*Sperophilus columbianus*). *Comp. Biochem. Physiol.* **41A**, 327–337.

Mullen, R. K. (1970). Respiratory metabolism and body water turnover rates of *Perognathus formosus* in its natural environment. *Comp. Biochem. Physiol.* **32**, 259–265.

Mullen, R. K. (1971). Energy metabolism and body water turnover rates of two species of free-living kangaroo rats. *Dipodomys merriami* and *Dipodomys microps. Comp. Biochem. Physiol.* **39A**, 379–390.

Nungesser, W. C. and Pfeiffer, E. W. (1965). Water balance and maximum concentrating capacity in the primitive rodent, *Aplodontia rufa. Comp. Biochem. Physiol.* **14**, 289–297.

Passmore, J. C., Pfeiffer, E. W. and Templeton, J. R. (1975). Urea excretion in the hibernating Columbian ground squirrel (*Spermophilus columbianus*). *J. exp. Zool.* **192**, 83–86.

Pennell, J. P., Lacy, F. B. and Jamison, R. L. (1974). An *in vivo* study of the concentrating process in the descending limb of Henle's loop. *Kidney Int.* **5**, 337–347.

Pennell, J. P., Sanjana, V., Frey, N. R. and Jamison, R. L. (1975). The effect of urea infusion on the urinary concentrating mechanism in protein-depleted rats. *J. clin. Invest.* **55**, 399–409.

Pfeiffer, E. W. (1968). Comparative anatomical observations of the mammalian renal pelvis and medulla. *J. Anat.* **102**, 321–331.

Pfeiffer, E. W. (1970). Ecological and anatomical factors affecting the gradient of urea and non-urea solutes in mammalian kidneys. In "Urea and the Kidney" (B. Schmidt-Nielsen and D. W. S. Kerr, eds). Excerpta Medica Foundation, Amsterdam, pp. 358–365.

Pfeiffer, E. W., Nungesser, W. C., Iverson, D. A. and Wallerius, J. F. (1969). The renal anatomy of the primitive rodent, *Aplodontia rufa*, and a consideration of its functional significance. *Anat. Rec.* **137**, 227–232.

Prong, T. A., Bjoraker, D. G. and Harvey, R. B. (1969). Comparison of the renal medullary vascular systems of dog and chinchilla. *Microvasc. Res.* **1**, 275–286.

Raab, J. and Schmidt-Nielsen, K. (1971). Effect of activity on water balance of rodents. *Fedn Proc.* **30**, 371.

Ramsay, J. A. (1935). Methods of measuring the evaporation of water from animals. *J. exp. Biol.* **12**, 355–372.

Raun, G. G. (1966). A population of woodrats (*Neotoma micropus*) in southern Texas. *Bull. Texas meml Mus.* **11**, 1–62.

Reynolds, H. G. and Turkowski, F. (1972). Reproductive variations in the round-tailed ground squirrel as related to winter rainfall. *J. Mammal.* **53**, 893–898.

Riad, Z. M. (1960). Pulmonary peculiarities to life in the desert. *Anat. Rec.* **137**, 99–103.

Richmond, C. R., Langham, W. H. and Trujillo, T. T. (1962). Comparative metabolism of tritiated water by mammals. *J. cell. comp. Physiol.* **59**, 45–53.

Riegel, J. A. (1972). "Comparative Physiology of Renal Excretion." Oliver and Boyd, Edinburgh.

Rocha, A. S. and Kokko, J. P. (1973). Sodium chloride and water transport in the medullary thick ascending limb of Henle. *J. clin. Invest.* **52**, 612–623.

Rodland, K. D. and Hainsworth, F. R. (1974). Evaporative water loss and tissue dehydration of hamsters in the heat. *Comp. Biochem. Physiol.* **49A**, 331–345.

Rothstein, A., Adolph, E. F. and Wills, J. H. (1947). Voluntary dehydration. In "Physiology of Man in the Desert" (E. F. Adolph and Associates, eds). Interscience Publishers, New York and London, pp. 254–270.

Rouffignac, C. de (1972). Physiological role of the loop of Henle in urinary concentration. *Kidney Int.* **2**, 297–303.

Rouffignac, C. de and Morel, F. (1966). Etude comparée du renouvellement de l'eau chez quatre espéces de rongeurs, dont deux especés d'habitat désertique. *J. Physiol., Paris* **58**, 309–322.

Rouffignac, C. de and Morel, F. (1969). Micropuncture study of water, electrolytes, and urea movements along the loops of Henle in *Psammomys*. *J. clin. Invest.* **48**, 474–486.

Rouffignac, C. de, Lechène, D., Guinnebault, M. and Morel, F. (1969). Etude par microponction de l'éboration de l'urine III. Chez le mérion non diurétique et du diurése par le mannitol. *Nephron* **6**, 643–666.

Rouffignac, C. de, Morel, F., Moss, N. and Roinel, N. (1973). Micropuncture study of water and electrolyte movements along the loop of Henle in *Psammomys* with special reference to magnesium, calcium and phosphorus. *Pflügers Arch. Eur. J. Physiol.* **344**, 309–326.

Rouiller, C. (1969). General anatomy and histology of the kidney. In "The Kidney—Morphology, Biochemistry, Physiology" (C. Rouiller and A. F. Muller, eds) Vol. I. Academic Press, New York and London, pp. 61–156.

Sachot, N., Sternberg, M., Skwarlo, K., Rabeyrotte, P. and Lagrue, G. (1975). Comparison of isolation and chemical composition of kidney glomerular basement membranes in rabbit, rat and man. *Comp. Biochem. Physiol.* **50A**, 575–579.

Schmid, W. D. (1972). Nocturnalism and variance in ambient vapour pressure of water. *Physiol. Zoöl.* **45**, 302–309.

Schmidt-Nielsen, B. and Kerr, D. W. S. (1970). "Urea and the Kidney." Excerpta Medica Foundation, Amsterdam.

Schmidt-Nielsen, B. and O'Dell, R. (1961). Structure and concentrating mechanism in the mammalian kidney. *Am. J. Physiol.* **200**, 1119–1124.

Schmidt-Nielsen, B. and Pfeiffer, E. W. (1970). Urea and urinary concentrating ability in the mountain beaver, *Aplodontia rufa*, *Am. J. Physiol.* **218**, 1370–1375.

Schmidt-Nielsen, B. and Schmidt-Nielsen, K. (1950a). Pulmonary water loss in desert rodents. *Am. J. Physiol.* **162**, 31–36.

Schmidt-Nielsen, B. and Schmidt-Nielsen, K. (1950b). Evaporative water loss in desert rodents in their natural habitat. *Ecology* **31**, 75–85.

Schmidt-Nielsen, B. and Schmidt-Nielsen, K. (1951). A complete account of the water metabolism in kangaroo rats and an experimental verification. *J. cell. comp. Physiol.* **38**, 165–181.

Schmidt-Nielsen, B., O'Dell, R. and Osaki, H. (1961). Interdependence of urea and electrolytes in production of a concentrated urine. *Am. J. Physiol.* **200**, 1125–1132.

Schmidt-Nielsen, K. (1963). "Osmotic regulation in higher vertebrates." Harvey Lectures, Series 58, pp. 53–95.

Schmidt-Nielsen, K. (1964). "Desert Animals. Physiological Problems of Heat and Water." Clarendon Press, Oxford.

Schmidt-Nielsen, K. (1969). The neglected interface: the biology of water as liquid-gas system. *Q. Rev. Biophys.* **2**, 283–304.

Schmidt-Nielsen, K. (1972). "How animals work." Cambridge University Press, Cambridge.

Schmidt-Nielsen, K. (1975). "Animal Physiology—Adaptation and Environment." Cambridge University Press, Cambridge.

Schmidt-Nielsen, K. and Haines, H. (1964). Water balance in a carnivorous desert rodent the grasshopper mouse. *Physiol. Zoöl.* **37**, 259–265.

Schmidt-Nielsen, K. and Schmidt-Nielsen, B. (1952). Water metabolism of desert mammals. *Physiol. Rev.* **32**, 135–166.

Schmidt-Nielsen, K., Hainsworth, F. R. and Murrish, D. E. (1970). Counter-current heat exchange in the respiratory passages: Effect on water and heat balance. *Resp. Physiol.* **9**, 263–276.

Shkolnik, A. and Borut, A. (1969). Temperature and water relations in two species of spiny mice (*Acomys*). *J. Mammal.* **50**, 245–255.

Silverstein, E. (1961). Urine specific gravity and osmolality in inbred strains of mice. *J. appl. Physiol.* **16**, 194–196.

Simon, G. T. and Chatelanat, F. (1969). Ultrastructure of the normal and pathological glomerulus. In "The Kidney—Morphology, Biochemistry, Physiology" (C. Rouiller and A. F. Muller, eds) Vol. I. Academic Press, New York and London, pp. 261–349.

Smith, B. W. and McManus, J. J. (1975). The effects of litter size on the bioenergetics and water requirements of lactating *Mus musculus*. *Comp Biochem. Physiol.* **51A**, 111–115.

Soholt, L. F. (1975). Water balance of Merriam's kangaroo rat, *Dipodomys merriami*, during cold exposure. *Comp. Biochem. Physiol.* **51A**, 369–372.

Sokolov, W. (1962). Skin adaptations of some rodents to life in the desert. *Nature, Lond.* **193**, 823–825.

Sperber, I. (1944). Studies of the mammalian kidney. *Zool. Bidr. Upps.* **22**, 249–432.

Spinelli, F., Wirz, H., Brücher, C. and Pehling, G. (1972). "Fine Structure of the Kidney Revealed by Scanning Electronmicroscopy." Ciba-Geigy, Basle.
Staaland, H. (1974). Absorption of mineral ions and water from the rabbit colon. *Norw. J. Zool.* **22**, 279–285.
Staaland, H. (1975). Absorption of sodium, potassium and water in the colon of the Norway lemming *Lemmus lemmus* (L.). *Comp. Biochem. Physiol.* **52A**, 77–00.
Steinhausen, M., Eisenbach, G. M. and Galaske, R. (1970a). Countercurrent system in the renal cortex of rats. *Science* **167**, 1631–1633.
Steinhausen, M., Eisenbach, G. M. and Galaske, R. (1970b). A counter-current system of the surface of the renal cortex of rats. *Pflügers Arch. ges. Physiol.* **318**, 244–258.
Stephenson, J. L. (1972). Concentration of urine in a central core model of the renal counterflow system. *Kidney Int.* **2**, 85–94.
Strömme, S. B., Maggert, J. E. and Scholander, P. F. (1969). Interstitial fluid pressure in terrestrial and semiterrestrial animals. *J. appl. Physiol.* **27**, 123–126.
Sullivan, L. P. (1974). "Physiology of the Kidney." Lea and Febiger, Philadelphia.
Tamarin, R. H. and Malecha, S. R. (1972). Reproductive parameters in *Rattus rattus* and *Rattus exulans* of Hawaii, 1968 to 1970. *J. Mammal.* **53**, 513–528.
Tennent, D. M. (1946). A study of the water losses through the skin of the rat. *Am. J. Physiol.* **145**, 436–440.
Thompson, R. (1971). The water consumption and drinking habits of a few species and strains of laboratory animals. *J. Inst. Anim. Techns.* **22**, 29–36.
Thurau, K. (1974). "Kidney and Urinary Tract Physiology" MTP Int. Rev. Sci., Physiology, Series one, Vol. VI. Butterworth, London.
Truniger, B. and Schmidt-Nielsen, B. (1970). Intrarenal distribution and transtubular movement of urea and related compounds. *In* "Urea and the Kidney" (B. Schmidt-Nielsen and D. W. S. Kerr, eds). Excerpta Medica Foundation, Amsterdam, pp. 314–322.
Tucker, V. A. (1965). Oxygen consumption, thermal conductance, and torpor in the California pocket mouse *Perognathus californicus*. *J. cell. comp. Physiol.* **65**, 393–404.
Valtin, H. (1973). "Renal Function: Mechanisms Preserving Fluid and Solute Balance in Health." Little, Brown and Co., Boston.
Vanjonack, W. J., Scott, I.M., Yousef, M. K. and Johnson, H. D. (1975). Corticosterone plasma levels in desert rodents. *Comp. Biochem. Physiol.* **51A**, 17–20.
Walker, E. P. (1964). "Mammals of the World" Vol II. The John Hopkins Press, Baltimore.
Weisser, F., Lacy, F. B., Weber, H. and Jamison, R. L. (1970). Renal function in the chinchilla. *Am. J. Physiol.* **219**, 1706–1730.
Whitaker, J. O. (1966). Food of *Mus musculus*, *Peromyscus maniculatus bairdi* and *Peromyscus leucopus* in Vigo county, Indiana. *J. Mammal.* **47**, 473–486.
Whitford, W. G. and Conley, M. I. (1971). Oxygen consumption and water metabolism in a carnivorous mouse. *Comp. Biochem. Physiol.* **40A**, 797–803.
Winkelmann, J. R. and Getz, L. L. (1962). Water balance in the Mongolian gerbil. *J. Mammal.* **43**, 150–154.
Wunder, B. A. (1970). Temperature regulation and the effects of water restriction on Merriam's chipmunk. *Eutamias merriami*. *Comp. Biochem. Physiol.* **33**, 385–403.
Yousef, M. K. and Bradley, W. G. (1971). Physiological and ecological studies on *Citellus lateralis*. *Comp. Biochem. Physiol.* **39A**, 671–682.
Yousef, M. K., Johnson, H. D., Bradley, W. G. and Seif, S. M. (1974). Tritiated water-turnover rate in rodents: desert and mountain. *Physiol. Zoöl.* **47**, 153–162.

Zahn, T. J. (1968). The effect of protein intake on maximum urine and renal tissue water-solute concentrations in the muskrat (*Ondatra zibethica osoyoosensis*, Lord). *Comp. Biochem. Physiol.* **25**, 1021–1033.

Zimny, M. L. (1968). Glomerular ultrastructure in kidneys from some northern mammals. *Comp. Biochem. Physiol.* **27**, 859–863.

Zimny, M. L., Hollier, L. and Clement, R. (1971). Ultrastructure of the proximal convoluted tubule of a hibernator correlated with renal enzymology. *Comp. Biochem. Physiol.* **40**A, 405–414.

4. Carnivores

R. L. MALVIN

University of Michigan, Michigan, USA

I.	Introduction	145
	A. General Problem of Osmoregulation	146
	B. Terrestrial Carnivores	147
II.	Central Osmoreceptors	148
III.	Hepatic Osmoreceptors	150
IV.	Low-pressure Receptors	153
V.	High-pressure Receptors	156
VI.	Angiotensin and ADH	157
VII.	Thirst	165
VIII.	Other Hormones	169
	A. Prostaglandin E	169
	B. Alpha and Beta Receptors	169
	C. Kallikrein	170
	D. Oxytocin	171
IX.	Summary and Conclusions	171
X.	Marine Carnivores	173
	A. Renal Function	174
	B. Antidiuretic Hormone	176
	C. Renin–Aldosterone	177
References		178

I. INTRODUCTION

This chapter will concern itself with a discussion of the control of osmoregulation in selected species of mammals, all of which include animal food as part of their diet. The term carnivores will not be taken in the strictest sense of including only those animals in the family of Carnivora, but rather include animals such as the rat, cat, dog, seal and porpoise. The reason for this is two-fold. First, those animals which include flesh as a part of their diet, be it a dog or a rat, have similar problems in terms of osmoregulation, and these problems are different from those of herbivores. Secondly, limiting the discussion to only those animals in the order Carnivora would severely limit the scope of this chapter. Considerable data have been accumulated from animals, such as the rat, which relate to all animals whose eating habits may be defined as carnivorous or omnivorous. The only distinction which will be made in this section will be between osmoregulation in mammals whose habitat is terrestrial and those which spend their lives in a marine environment. As is evident, the environment of marine flesh eaters is so different from that of the land animals that their problems of osmoregulation are unique and deserve special consideration.

Finally, a word should be said about the meaning of osmoregulation as it is taken in this chapter. In general I will be concerned with those control systems which serve to maintain constant the osmolality and the concentration of Na in plasma. No doubt, this is a restricted definition of osmoregulation as there are many additional control systems which serve to regulate the concentrations of solutes other than Na. But, Na is the major extracellular ion and is traditionally thought of as the main solute which determines directly the osmolality of the extracellular compartment. I will also include some discussion of the control of extracellular volume as the connections between the body's control of volume and osmolality are so intertwined as to require consideration of both.

A. GENERAL PROBLEM OF OSMOREGULATION

When life began in the primordial sea, the problem of osmoregulation could hardly have existed. A single cell living in an environment very similar to its own internal environment would encounter minimal difficulties maintaining its osmolality constant. Simple diffusion of solutes across the cell surface would be sufficient to maintain the constancy of the internal environment required for life. However, as multicellular organisms developed, so developed special problems of maintaining cellular osmolality. Diffusion alone could no longer serve the animal's needs; as the length of the diffusion path to the sea became too great, and so a circulatory system was necessary. In addition, the multicellular organism requires the existence of controls which serve to maintain the volume of the circulatory system constant. As the primordial sea slowly evolved into the more concentrated one of today, and as animals left the sea and entered the land, additional problems of survival were imposed upon the evolving species. Among them was the fact that access to water was intermittent, so that osmolality had to be preserved during periods of water deprivation. In addition, many animals had superimposed periods of Na surfeit and deficit. Other species were faced with the problem of using water to maintain body temperature constant in environments in which ambient temperature and humidity might vary almost to the limits of endurance. Thus, for successful adaptation, species needed to develop mechanisms not only for conserving water and Na in times of deficit, but also to acquire mechanisms to drive the animal to resupply those elements when depleted, i.e. thirst for water and appetite for Na. An attempt will be made to discuss those systems which control intake and output of Na and water, and finally to attempt to put them together in an integrated fashion.

B. TERRESTRIAL CARNIVORES

1. *General Considerations*

All multicellular animals are faced with similar problems of survival, it is only the magnitude of those problems which differ. For any individual to survive, it must maintain a constant concentration of electrolytes in its body fluids; a change of only a few per cent often results in a serious dislocation of the animal's well being. It must also prevent the total osmolality of the body fluids from deviating significantly from some mean value. Finally, even if electrolyte concentrations and osmolality are maintained constant, the total volume of the body compartments, particularly the plasma volume, must also be kept constant.

The difficulty imposed upon the individual animal in solving these homeostatic problems is dependent in large part on the access each animal has to a supply of the necessary salts and water. This is not to say that one need only consider how it is that an animal is driven to consume salt and water, to understand how it survives. That clearly is not the case. But equally clear is the fact that, in regards to water balance, the camel and the beaver have very different levels of difficulty in surviving. Yet, even in this narrowly defined problem area they do have one thing in common. Both animals presumably experience thirst, and both drink water to satisfy that thirst.

A significant part of this section will be devoted to a description of those control systems which serve to keep animals in Na balance. Again, the severity of this problem varies somewhat from species to species. In general, herbivores are more prone to suffer long periods of severe Na restriction (Denton, 1972). Carnivores, on the other hand, are seldom forced to face extended periods of Na deprivation. The very nature of their diet prevents that. Again, it is only the magnitude of the problem which distinguishes herbivores from carnivores. Flesh eaters do have to adjust the output of salt to equal a variable intake. It is only the lower limit of adjustment which differs between the two groups.

2. *Water Loss*

Although terrestrial mammals lose water through a variety of routes, only one (urinary loss) is controlled by systems which have as their function the maintenance of plasma osmolality or volume. Water lost through the respiratory tract varies with the ambient temperature and humidity and is not controlled by the internal water balance of the body. Similarly, loss of water and Na in sweat is variable, but depends upon considerations of temperature and activity rather than the state of water or salt balance. Minor volumes of water are lost in tears, saliva and faeces, but, again, there is minimal control over the volumes lost via these routes. The only important route of water loss, in terms of the control of osmoregulation, is by way of urine. The limits of this loss vary between as little as 0·5% to as much as 20% of the glomerular filtration rate. Of all the factors known to affect the rate of urine flow antidiuretic hormone (ADH) is, by far, the most potent. In fact, ADH appears to be the only substance capable of affecting only water transport. Other hormones or physical factors known to alter urine flow cause changes in the excretion rate of Na and other urinary solutes.

3. *Sodium Loss*

The regulation of Na loss is almost entirely confined to the kidney. Some Na is lost in the faeces, but this is an essentially constant quantity. The kidney, on the other hand, is able to alter Na excretion by a factor of many hundreds. By far the most important single factor involved in the regulation of urinary Na excretion is the hormone aldosterone which promotes reabsorption of Na from the filtrate. It is the regulation of aldosterone blood levels which is complex and involves many afferent receptors.

II. CENTRAL OSMORECEPTORS

Since the classical paper of Verney (1947) it has been accepted that osmoreceptors exist somewhere within the brain, presumably within the hypothalamic regions. It was Verney who first demonstrated that water diuresis could be inhibited by the injection of impermeant hypertonic solutes. Since that time, considerable additional work has been done in locating and in characterizing those osmoreceptors. Brooks *et al.* (1966) and Hayward and Vincent (1970) demonstrated an increase in the firing rate of supraoptic and paraventricular cells on infusion of hypertonic solution in the carotid artery. Emmers (1973) was able to record action potentials from the supraoptic and paraventricular nuclei of the cat and, in addition, from the area lateralis hypothalami and nucleus entopeduncularis. He showed, in a very nice manner, that the intracarotid infusion of 3% NaCl increased spiking from both the supraoptic and paraventricular nuclei, but not from the other areas studied. However, he also pointed out that activity of both the paraventricular and supraoptic nuclei may be modified by activity in other nuclei; that is, there are both inhibitory and excitatory influences playing upon the paraventricular and supraoptic nuclei. Similar results were previously reported by Koizumi *et al.* (1964) who demonstrated the presence of inhibitory input from the cerebellum. Injections of adrenaline were also inhibitory to those neurons. Similar data have also been recorded by others which show that the spike activity from the supraoptic nucleus is increased very significantly during periods of water deprivation, but that activity of the surrounding areas is not changed to any significant degree (Walters and Hatton, 1974).

Although it has been amply demonstrated that spike activity increases in those nuclei known to be associated with the secretion of ADH when animals are subjected to hypertonic stimuli, the relative importance of those receptors in the control of ADH secretion is another problem. Dunn *et al.* (1973) did attempt to get at precisely this problem. In Fig. 1 are shown the results of a series of experiments in which they were able to show that osmoreceptors are of major importance in the regulation of ADH secretion. They gave saline, hypertonic saline or polyethylene glycol intraperitoneally to rats and followed the changes in blood volume, osmotic pressure of plasma and the concentration of ADH in plasma. From Fig. 1 it is evident that the increase in ADH concentration of blood rose most steeply as plasma osmolality was changed. Similar proportional changes in plasma volume, although they did serve to increase ADH levels, were far less effective than a similar change in plasma osmolality.

Thus, it appears fair to say that cells which are able to respond to the osmolality of plasma are located in both the supraoptic and paraventricular nuclei. These cells are able to signal the posterior pituitary to release ADH in response to those stimuli. Furthermore, it appears that the hypothalamic nuclei may also be inhibited by other brain

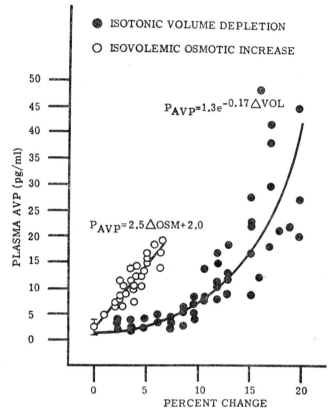

Fig. 1. The relationship of plasma AVP (ADH) to percentage change in plasma osmolality and blood volume. The percentage change in osmolality (OSM) was calculated relative to the mean basal level of 294 mosmol kg^{-1}. The percentage change in blood volume (VOL) was calculated relative to mean basal haematocrit. Each point represents one animal (from Dunn et al., 1973).

areas yet to be described, and that osmoreceptors seem to be of vital importance in regulating ADH secretion. At least in a proportional way they appear to be 2 to 3 times more effective in regulating ADH levels than are changes in blood volume.

III. HEPATIC OSMORECEPTORS

Haberich et al. (1964) first reported that there appear to be osmoreceptors located somewhere within the hepatic circulation. They were able to show that rats when administered water through the hepatic portal vein had a much larger diuretic response to the water load than if the same volume were administered via a systemic vein. These original observations have been repeated and extended by Haberich (1971) and others, most of whom reported that infusions of either hypotonic or hypertonic solutions into the portal vein were more effective in causing a diuresis and a natriuresis than similar infusions given via a systemic vein or directly into the vena cava.

Dennhardt et al. (1971) were able to extend the original observations using a rather ingenious experimental model. They prepared rats with catheters placed in the hepatic portal vein and in the vena cava. They then infused water through one catheter and 1·8% NaCl through the other. Both solutions were infused at the same rate, thus the mixture was essentially an isotonic solution. When the saline solution was infused into the portal vein and water into the vena cava, no significant diuresis resulted. However, when the infusions were reversed, and the portal vein received water while the vena cava received the infusion of 1·8% NaCl, a larger and significant diuresis occurred. Even more significant was the finding that if the vagus nerve was severed, the expected diuresis did not occur. The implication of these experiments is that there are osmoreceptors located in the rat liver which have their afferent path in the vagus nerve.

Presumably the vagal afferents end somewhere in the hypothalamus. This hypothalamic centre in turn must control the rate of release of a hormone which affects the rate of urine flow. These experiments however, do not give a clue as to whether or not that hormone is ADH or some other diuretic substance, or even an antidiuretic substance whose secretion is inhibited by a low hepatic blood osmolality.

Experiments similar to these described above were done by a host of other workers who infused hypertonic saline into the portal vein and compared the response to that infusion with a similar infusion into a peripheral vein. Standhoy and Williamson (1970) injected hypertonic saline into the hepatic vein and were able to show that there was a greater increase in Na excretion when the solution was infused directly into the hepatic vein than when given to the animal via the femoral vein. Passo et al. (1972) did similar experiments on cats in which they infused 1M NaCl at 0·2 ml min^{-1} into the portal vein or into the femoral vein. Femoral vein infusion had little or no effect during the short time of the experiment, while the infusions into the portal vein increased Na excretion three-fold and urine flow by 50%. More significantly, if the animals, in this case, cats, were vagotomized hypertonic infusions of saline into the portal vein had no effect on Na excretion or urine flow rate. Similar experiments were done by Daly et al. (1967) in dogs first prepared so that they were given maximal doses of ADH and 9-alpha-fluorohydrocortisone. The dogs were then infused with a 5% NaCl solution at 0·07 ml kg^{-1} min^{-1} either into the portal vein or the femoral vein. Portal infusions resulted in a significantly greater increase in urine flow rate and Na excretion than infusions given via the femoral vein. These workers were able to show that neither infusion caused a significant change in the glomerular filtration rate. It is hard not to interpret these experiments as evidence for some natriuretic factor being released centrally due to portal stimulation or an antidiuretic factor being inhibited. That the liver contains a significant number of receptors which might mediate the response has been shown by Andrews and Orbach (1973). Monitoring impulses from the hepatic nerves, they were able to demonstrate that a large variety of substances either increased the number of action potentials along those nerves or inhibited them. For example, acetylcholine, 5-hydroxytryptamine, bradykinin and epinephrine all increased the rate of discharge along the hepatic nerves. On the other hand, pentolinium, hexamethonium and aspirin significantly inhibited the background rate of firing. Similar results have been reported when nerve impulses in the hepatic vagus of perfused rat livers was monitored (Akachi et al., 1976). They were able to demonstrate that an increase in the concentration of Na—as little as 5 mM in the perfusate—was sufficient to stimulate increased nerve activity.

All of the above data are consistent with the hypothesis that osmoreceptors are located within the liver. However, there also exists a body of data which are in direct contradiction to the results thus far reported. For example, Schneider et al. (1970) infused water into the portal vein or the vena cava of dogs. In their hands, the animals showed an identical response to the infusions, regardless of the route of administration. They were unable to detect any difference in response between infusions made directly into the liver or into the vena cava. They also repeated the split-stream experiments. Regardless of whether the hypertonic saline or water was given to the liver or the peripheral vein, the response was the same. They were unable to find any evidence for the existence of hepatic osmoreceptors or Na receptors. Potkay and Gilmore (1970), using anaesthetized dogs, were also unable to find any difference in response between dogs given 5% NaCl solution directly into the portal vein at 0·75 ml kg^{-1} min^{-1} or given the same dose into a systemic vein. Glasby and Ramsay (1974) infused 0·45% NaCl into either the hepatic portal vein or a systemic vein. In contrast to similar experiments by Lydtin (1969a, 1969b) they were unable to determine any difference in $C_{osm'}$, $C_{H_2O'}$, $U_{Na}V$ or U_KV. Perhaps it should also be mentioned here that Haberich himself has reported evidence that is contrary to the existence of some type of Na receptor within the hepatic circulation (Haberich, 1968). He reported that an injection of hypertonic saline into the portal vein resulted in a decrease in urine flow rather than the increase reported by other workers.

Precisely why such conflicts in the literature exist is rather difficult to understand. There does not appear to be any consistent pattern which depends upon species used—dog, cat or rat—nor does the presence or absence of anaesthesia appear to be a consistent finding in whether or not one finds evidence for hepatic osmoreceptors, neither does the dose of injected water or hypertonic saline appear to be a determining factor. In fact, it appears that identical experiments done in different laboratories yield diametrically opposed results.

One is quite hard pressed to determine whether or not one group finds osmoreceptors due to some artefact of the experiment or whether the opposing group is unable to find them because of some inappropriate methodology which escapes detection. I believe that it is safe to say, at this stage, that the problem is not solved, but that the likelihood does exist that the liver contains either an osmoreceptor or a Na receptor which is capable of altering urine flow rate and Na excretion. It is worth noting that Andrews and Orbach (1974) were able to show that changes in the Na concentration of as little as 10 mmol perfusing the liver caused a significant change in the discharge rate of nerves leaving the liver.

IV. LOW-PRESSURE RECEPTORS

In 1954, Gauer *et al.* published a series of experiments which initiated a new field of research in renal physiology. In this classical paper, it was shown that urine flow increased significantly in animals which were subjected to negative pressure breathing. The increase in flow rate of urine was reversed when negative pressure breathing ceased. By measuring cardiac output, mean blood pressure and other haemodynamic parameters, they were able to suggest that somewhere within the cardiovascular system, probably in the pulmonary circulation or in the chambers of the heart, there exist stretch receptors, such that when the volume of blood in the chest is increased, stretch of these receptors inhibited the secretion of ADH and diuresis ensued.

Since that time many workers have been able to extend these observations and to show rather conclusively that there are indeed stretch or pressure receptors in the low-pressure side of the cardiovascular system which control the rate of secretion of ADH. Henry and Pearce (1956) were able to increase left atrial pressure by placing a small balloon within the atrium and inflating it at will. As the balloon diameter increased, the left atrial pressure rose. They found that as atrial pressure increased, so did urine flow. It was hypothesized that these atrial receptors signalled the central nervous system by way of the fibres originally described by Nonendez (1937). Johnson et al. (1969) carried these observations somewhat further. They, too, implanted a balloon in the left atrium, inflated the balloon, and measured left atrial pressure. However, in addition, they also measured the concentration of ADH in plasma. In this way they were able to relate the circulating level of

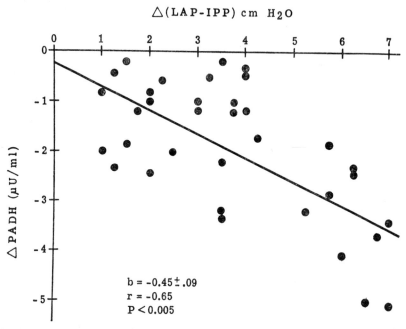

Fig. 2. Change in left atrial transmural pressure versus change in plasma ADH levels (from Johnson et al., 1969).

ADH to left atrial pressure. In Fig. 2 are shown the results of those experiments. It is evident that an excellent inverse relation exists between left atrial pressure and the ADH concentration of blood.

In addition to the presence of left atrial receptors, atrial receptors in the right heart have also been demonstrated. Brennan et al. (1971) increased right atrial pressure by inflating a balloon placed in the atria of anaesthetized dogs. They were able to show that increased right atrial pressure had no effect on ADH release. However, increases in right atrial pressure of only a few cm H_2O significantly reduced the plasma renin activity. Similar results were obtained by Anat et al. (1976). Stretching of the right atrium by traction on threads attached to the atrial wall reduced renin secretion. This effect was absent if the vagus nerves were sectioned. Thus both left and right atrial receptors exist, and both are involved with control of hormones which regulate salt and water balance. Stitzer and Malvin (1975) have demonstrated that the right atrium may contain stretch receptors which also serve to regulate Na excretion. Inflation of a right atrial balloon resulted in paradoxical Na retention which was unrelated to any change in GFR, RBF, filtration fraction or mean arterial pressure.

Although it appears clear that somewhere within the low-pressure system on the left side of the heart, receptors respond to stretch and so signal the change in release of ADH, there is some controversy in the literature regarding the nature of the effect and the possible efferent arc of that reflex. Ledsome and Mason (1972) distended the left atrium of dogs and observed an increase in urine flow rate and an increase in the free water clearance. The effect, however, was blocked only by very large doses of vasopressin, leading the authors to conclude that although ADH did play a role in the diuresis resulting from atrial distention, something else, neural or hormonal, was also involved. Lydtin and Hamilton (1964) arrived at similar conclusions a few years earlier. They placed a snare around the mitral valve which could be tightened at will; a procedure which caused an increase in atrial pressure. They were able to increase left atrial pressure by as much as 20 cm of H_2O, which also led to a diuresis and a decrease in the specific gravity of the urine. This response, however, was not inhibited completely by the infusion of 0·025 mU kg^{-1} min^{-1} of ADH, leading these authors to conclude as did Ledsome and Mason (1972) that the response was due to something other than ADH. Kappagoda *et al.* (1975) reported that left atrial distension in hypophysectomized dogs resulted in diuresis. These authors believe their data implicate a hormone other than ADH in the diuretic response to left atrial distention. These results are somewhat hard to reconcile with those of Kinney and DiScalla (1972) who placed a balloon in the left atrium, increased left atrial pressure and found that the diuresis which ensued was completely blocked by infusions of ADH. In addition, they were able to show that distal tubular blockade of Na transport with diuretics prevented the diuretic response to left atrial distention. This implies that left atrial distention has a distal effect on the tubule, which fits with the hypothesis that left atrial distention is responsible for decreasing the circulating level of ADH, and that ADH alone is responsible for the diuresis of atrial distention.

Although all workers who have increased left atrial pressure report that urine flow is increased (Henry and Pearce, 1956; Brennan *et al.*, 1971; Gillespie *et al.*, 1973), there is a controversy regarding the effect of left atrial pressure on the excretion of Na and K. Gillespie *et al.*

(1973) reported that increases in left atrial pressure of 5 to 10 cm H_2O increased both the excretion of Na and K and that this natriuresis and kaliuresis was independent of the GFR, RFP or the circulating level of ADH. They theorized that the increase in Na and K excretion could be due to some neural or hormonal changes. These results are in agreement with those of Kappagoda et al. (1973) who showed that atrial distention increased both urine flow and Na excretion very rapidly, within 5 to 10 min. This suggests that the result could not be due solely to a fall in the circulating level of ADH, which has a half-life far too long to allow changes to occur within 5 min.

Goetz et al. (1970) investigated this problem using a somewhat different and ingenious technique. They created a pericardial pouch around the atria of dogs and were able to change the atrial transmural pressure by injecting saline into the pouch. The pouch pressure was changed while the dogs were unanaesthetized and had recovered fully from the surgery. In their experiments, an increase in pericardial pressure, or a decrease in the transmural pressure, led not only to a fall in urine flow rate, but also to a decrease in Na excretion. These changes in urine flow rate and Na excretion were not accompanied by changes in creatinine clearance, aortic blood pressure or cardiac output.

Thus it appears that atrial receptors are capable not only of altering the flow rate of urine, presumably by changing the output of ADH, but also by some other as yet unknown mechanism, are able to alter the excretion rate of Na (Mulcahy et al., 1975). It is, of course, possible that some of the effect on Na excretion is due to changes in intra-renal distribution of blood flow. Kahl et al. (1974) showed that left atrial hypotension results in renal vasoconstriction and that atrial volume receptors appear to cause a decrease in renal blood flow, and presumably glomerular filtration rate, which in turn would cause an increase in water reabsorption.

In summary, there is significant evidence to document the hypothesis that atrial receptors exist which affect both the excretion of salt and water. Unquestionably, atrial receptors signal the pituitary to alter the rate of secretion of ADH. In addition, atrial receptors appear to alter the rate of Na excretion by the kidneys. Whether this latter effect is hormonal or neural remains to be seen, although a hormonal effector appears to be the more likely.

V. HIGH-PRESSURE RECEPTORS

In addition to afferent receptors responding to pressure or stretch in the venous system, there is ample evidence that receptors are also located in the high-pressure side of the circulatory system. Share and Levy (1962) were the first to show that if the vagus nerves were cut, the ADH titre of plasma increased. If the carotid arteries were occluded, with intact vagus nerves, no change was noted in the ADH levels of plasma. However, a combination of vagotomy and carotid occlusion resulted in an increased level of circulating ADH. Their conclusions were that baroreceptors, located in the carotid sinus, have a tonic inhibitory influence on ADH secretion. Additional evidence in both the dog and the cat has been gathered by Share (1965, 1967, 1968) and by Clark and Silva (1967) showing that high-pressure baroreceptors do indeed exist in the area of the carotid sinus in both dog and cat, and that the vagus and carotid sinus afferents are inhibitory to the secretion of ADH. Furthermore, it has also been shown that the carotid sinus is responsive to the pulse pressure rather than solely mean blood pressure (Share and Levy, 1966). If carotid sinus mean pressure was kept constant, but the pulse pressure reduced to zero, ADH increased. It should be noted, however, that the authors used only two pulse pressures, 0 and 90 mm Hg, neither of which are within the physiological range. There remains then, some question regarding the role of the pulse pressure in the normal day-to-day regulation of ADH secretion.

Thus, both sides of the circulatory system, the arterial and the venous, contain receptors which are able to signal the posterior pituitary as to the cardiovascular status. When pressure is high in either system, ADH secretion is reduced, and when pressure is low in either system, the secretion of ADH is stimulated. It is reassuring to know that this system makes teleological sense, in that it would serve to maintain blood volume at some constant level. Increases in volume would cause increased left atrial pressure as well as mean arterial blood pressure. ADH levels would then decrease with a resulting diuresis. This should aid in returning blood volume to normal.

VI. ANGIOTENSIN AND ADH

Although considerable evidence has been amassed in the literature that the renin–angiotensin system plays some role in the maintenance of blood pressure, it has only been in the past five years or so that its role in the regulation of ADH secretion has been demonstrated. Over a decade ago Dickinson and Lawrence (1963) showed that an infusion of angiotensin into the vertebral artery was more potent in raising blood pressure than a similar infusion given intravenously. Since that time, other authors have been able to show that angiotensin infused into the arterial system of the brain, given directly into the brain substance or in the cerebral spinal fluid causes an increase in mean blood pressure. That this response was not due solely to angiotensin getting into the general circulation and causing an increase in pressure by a direct effect on the peripheral vasculature was shown by experiments in which the hypertensive activity of the injected angiotensin was completely inhibited by cervical spinal section (Smookler et al., 1966; Severs et al., 1967). Even more to the point was the finding that the pressor activity is abolished in unanaesthetized hypophysectomized, ganglion blocked rats (Severs et al., 1970). These experiments indicate that angiotensin has some central action on blood pressure and that that action is dependent upon pituitary function.

TABLE I

Effect of intravenous infusion of angiotensin (10 ng kg^{-1} min^{-1}) on the concentration of ADH in the plasma of conscious dogs.

	Control	Angiotensin	Recovery
ADH, μU/ml	1·4 ± 0·2	2·3 ± 3^a	1·5 ± 0·1
MBP, mm Hg	118 ± 5	113 ± 6^a	119 ± 6
Posm	305 ± 2	305 ± 2	306 ± 2
[Na] plasma	152 ± 1	153 ± 1	151 ± 1

Values are means ± SE
Number of dogs is 6.

^{a} Difference significantly different from control, $P < 0.01$
From Bonjour and Malvin (1970).

It has long been known that the posterior pituitary gland secretes ADH which is capable of increasing blood pressure. Thus, these experiments tentatively suggest that angiotensin may in some way stimulate the pituitary to release ADH. Bonjour and Malvin (1970) were the first to measure changes in the ADH level of plasma in response to infusions of angiotensin. In Table I are shown the results of a series of experiments on unanaesthetized dogs. It can be seen from Table I that infusion of 10 ng kg^{-1} min^{-1} of angiotensin intravenously was sufficient to increase the ADH level of plasma approximately 65%. It may be worth noting, at this point, that a change in ADH concentration from 1·4 to 2·3 μU ml^{-1} in an unanaesthetized animal is sufficient to reduce urine flow approximately 60% (Zehr et al., 1969). Thus, in these experiments it was evident that a rather substantial change in physiological function is produced by the infusion of a dose of angiotensin sufficient to raise the mean blood pressure by only 15 mm Hg. It should also be noted that it was shown by Share (1965, 1968) that an increase in mean blood pressure is an inhibitory stimulus to the secretion of ADH. Thus, in the face of an inhibitory stimulus coming from the high-pressure system, angiotensin was still effective in eliciting an increase in ADH titre.

In Fig. 3 are shown the results of experiments in which different concentrations of angiotensin were infused into the ventricular-cysternal system of anaesthetized dogs. At all levels of angiotensin infusion, the circulating level of ADH was increased. In none of these experiments, however, was any change seen in blood pressure, the dose of angiotensin

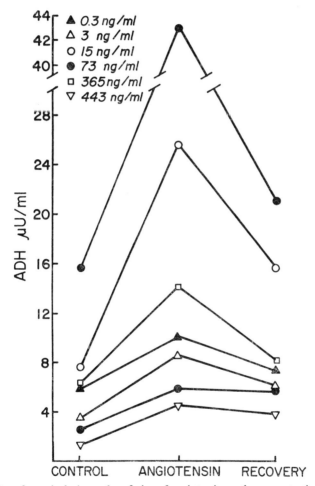

Fig. 3. Effect of ventriculocisternal perfusion of angiotensin on the concentration of ADH in peripheral plasma. Each series of three points represents results obtained from an experiment in which concentrations of angiotensin in CSF was as shown on figure (from Mouw et al., 1971).

being subpressor. In Fig. 4, these results are plotted somewhat differently. Along the abscissa is plotted the ADH concentration in $\mu U\ ml^{-1}$ during the control period of observation. Along the ordinate is plotted the change in ADH concentration seen after the infusion of angiotensin through the ventricular-cysternal system. It is evident from Fig. 4 that there is a direct relationship between the starting level of ADH and the

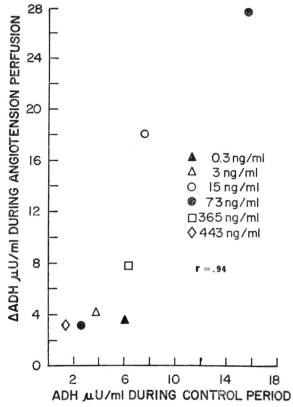

Fig. 4. Change in ADH concentration of plasma as a function of the ADH concentration in the control period. Plotted from data in Fig. 3 (from Mouw et al., 1971).

absolute increase in the circulating level occasioned by the infusion of angiotensin. From these data, it is suggested that a given stimulus alters the fractional rate of secretion rather than causing an absolute change in secretory rate. That is, when secretory rates are low, as judged by low plasma levels, a given stimulus will cause a relatively small increase in the secretory rate. When secretory rates are high, as judged by high levels of the circulating hormone, the same stimulus will result in a greater absolute rate of release. This interpretation is acceptable only if all the doses used in these experiments were maximal doses. This, in fact, appears to be the case, as each dose caused approximately the same percentage increase in plasma level of ADH (Malvin, 1972).

That the ADH titre of blood is increased following the administration of angiotensin II centrally, is further supported by the work of Severs et al. (1970, 1971). They were able to show that water-loaded rats given angiotensin II in the cerebrospinal fluid, not only drank more water than control rats, but showed higher urine osmolalities and lower urine volumes than control rats. In fact, the plasma actually became hypotonic while the rats produced hypertonic urine. These data are consistent with the hypothesis that angiotensin stimulated both thirst and the secretion of ADH, resulting in a decreased fluid output and hypoosmolality of the plasma. Similar experiments were done by Olsen (1974) showing that both intracarotid and intreventricular infusion of angiotensin resulted in both antidiuresis and increased urine osmolality. Although the authors did not measure ADH, the results are consistent with an increased circulating level of that hormone resulting from the angiotensin infusion. Similar experiments in humans have also been reported. Infusion of angiotensin II sufficient to raise blood pressure 20 mm Hg increased the ADH concentration of plasma two- to four-fold (Uhlich et al., 1974). Keil et al. (1975) report that if 0, 10, 50 or 100 ng of angiotensin was injected into the ventricular cerebrospinal fluid of conscious rats a good dose-response relation was obtained. In Fig. 5 are plotted the results of those experiments. As can be seen, each dose of angiotensin increased the circulating level of ADH significantly above resting levels.

Some direct evidence also exists concerning the receptor site for angiotensin. Nicoll and Barker (1971) were able to show that when angiotensin was administered in very small doses directly to the cells of the supraoptic nucleus, spike activity increased, while activity of cells in other brain areas were unaffected by the administration of angiotensin. Similar results were reported by Sakai et al. (1974). Explants of dog supraoptic nuclei were cultured for two–three weeks. Superfusion of the explants with fluid containing angiotensin caused a significant increase in spike activity.

Gregg and Malvin (1978) incubated either isolated posterior pituitaries of rats or the posterior pituitary still attached to the hypothalamus by the stalk. ADH secretion into the medium was measured before, during and following the addition of angiotensin II to the medium. Angiotensin II was found to have little or no effect on the isolated pituitaries. However, a 500% increase in ADH secretion was obtained from the intact hypothalamic pituitary system. These data are, of course, consistent with the hypothesis that angiotensin-stimulated

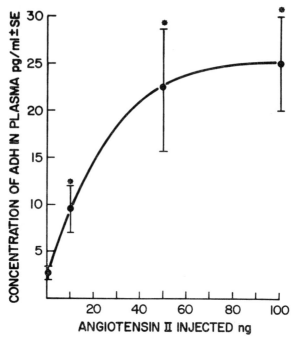

Fig. 5. The concentration of ADH in plasma as a function of the angiotensin (in ng) injected into the ventricular cerebrospinal fluid (plotted from data in Keil *et al.*, 1975).

hypothalamic nuclei which, in turn, stimulated the posterior pituitary to release ADH. Again, these data are consistent with those of Nicoll and Barker (1971) and of Sakai *et al.* (1974) indicating that the receptor site for angiotensin is in the hypothalamic nuclei.

Although most workers appear to agree that angiotensin is able to affect the secretion rate of ADH, there is a small apparent disagreement in the literature. Claybaugh *et al.* (1972) using anaesthetized dogs, infused angiotensin intravenously over a dose range from 10 to 60 ng kg^{-1} min^{-1}. They reported that they were unable to increase the plasma level of ADH by such infusions. However, it should be noted that in a later paper these same workers (Shimizu *et al.*, 1973) showed data which clearly implied that angiotensin is able to stimulate the release of ADH. Following water loading in dogs, they infused either hypertonic saline alone or hypertonic saline plus angiotensin. At essentially all plasma osmolalities studied, the response to angiotensin and saline was much greater than that with hypertonic saline alone, indicating that angiotensin does indeed have a central effect. At plasma osmolalities of 325 mosmol kg^{-1} H_2O, infusion of angiotensin at 10 ng kg^{-1} min^{-1} increased ADH levels approximately $2\frac{1}{2}$-fold. More recently

Claybaugh (Claybaugh, 1976) demonstrated that physiological levels of renin stimulate ADH secretion in the dehydrated conscious dog, but not in the hydrated dog. Precisely how this release is mediated is another problem, however. Whether the release results from direct stimulation of receptor sites causing hypothalamic neurons to fire to the pituitary or the stimulatory effect is secondary to some membrane change affected by angiotensin is, as yet, unknown. But in either case, the physiological function of angiotensin is the same. Increased levels of circulating angiotensin will result in increased release rates of ADH.

Not only has it been shown that angiotensin stimulates ADH release, but a negative feedback control of that system has also been demonstrated. It is well documented now that ADH, acting directly upon the kidney, is able to inhibit the secretion of renin, which in turn, of course, generates angiotensin (Bunag et al., 1967; Vander, 1968; Tagawa et al., 1971). Thus, we have the full loop of a negative feedback system—angiotensin stimulates ADH release, ADH in turn inhibits renin release.

It has also been known for some years, due to the elegant work of Laragh et al. (1960) that a major function of angiotensin is in the stimulation of aldosterone secretion, which in turn causes the kidney to increase the reabsorptive rate of Na. Thus, angiotensin has the effect of aiding the body in saving both salt and water. Furthermore, the work of Oelkers et al. (1974) indicates that not only does angiotensin stimulate the secretion of aldosterone, but the actual Na state of the animal is a determinant of the effect that a given dose of angiotensin will have. In Fig. 6 are shown the results of a series of their experiments done on human subjects who were either depleted of Na by three days of a low Na diet plus 30 to 60 mg of furosemide given intravenously or a series of subjects on a normal Na diet. At all levels of angiotensin infusion, aldosterone secretion was increased, but note that the slope relating angiotensin infusion to aldosterone concentration of plasma is steeper in the Na-deplete subjects than in the Na-replete subjects, indicating that Na depletion in some ways sensitizes the adrenal glands to angiotensin.

Angiotensin also appears to stimulate the release of anterior pituitary hormones, which in turn stimulate aldosterone release (Daniels-Severs et al., 1971; Mulrow and Ganong, 1961). Further, it is now known that aldosterone, in addition to its Na-retaining effect on the kidney also stimulates salt appetite in animals (Fitzsimons, 1971). Angiotensin, then, is directly and indirectly responsible for altering the rate of release of aldosterone by the adrenal gland, and also in controlling, in part, the intake of Na.

In addition to its ability to stimulate ADH and aldosterone secretion, angiotensin has a more direct action on renal Na transport. Barraclough et al. (1966) and Malvin and Vander (1967a) have shown that intravenous infusions of relatively small doses of angiotensin II decreases Na excretion. Although the reduction in $U_{Na}V$ in these experiments could be a result of the small decrease in GFR occasioned by the infusion, Johnson and Malvin (1975) demonstrated that physiological doses of angiotensin infused into one renal artery of a dog decreased $U_{Na}V$ with no change in GFR or RPF. Thus, angiotensin II

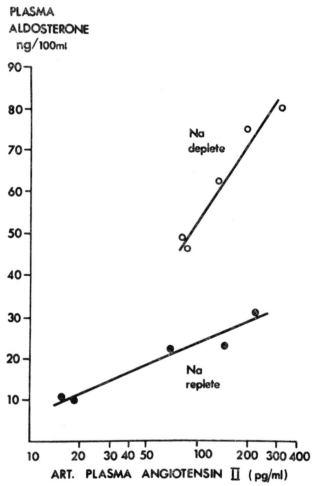

Fig. 6. Concurrent arterial plasma angiotensin II and plasma aldosterone concentrations before (solid circles) and after (open circles) sodium depletion in one subject. Abscissa is a log scale. The difference in slope of the two curves is highly significant on F-test ($p < 0.001$). This difference was also apparent in the zone of overlap ($p < 0.05$) (from Oelkers et al., 1974).

appears to stimulate tubular Na reabsorption and, in a sense, work synergistically with aldosterone in saving Na.

VII. THIRST

That animals have a very strong drive to replace water losses has certainly been known since the beginning of time. The problem in recent years has been to define those mechanisms responsible for that thirst drive. Clearly, one of the factors involved in thirst must be a sensing mechanism which is able to detect changes in plasma osmolality because it is this single parameter which alone can be controlled by the variations in intake and output of water. It makes teleological sense that the body should contain, somewhere, a receptor which is sensitive to the osmotic pressure of plasma or interstitial fluids and so signal the animal to either drink or to stop drinking water when the plasma osmolality moves from some mean value.

That there is a thirst centre within the brain was first shown by Greer (1955). He placed electrodes in the hypothalamic region of a rat. When stimulated, the rat exhibited a very strong drinking drive. This appears to be the first demonstration of a drinking centre. This centre is now known to be activated by a variety of agents, but among them, most certainly, is an alteration in the osmotic pressure of plasma. Measurement of the activity of single cells in the preoptic region (thirst centre?) indicates that activity increases when the cells are subjected to hypertonic solutions (Cross and Green, 1959). It has been demonstrated that when osmolality of plasma is increased by the infusion of hypertonic saline, drinking behaviour is initiated (Fitzsimons, 1961, 1963; Holmes and Gregerson, 1950a, 1950b). In fact, evidence indicated that not only is there an osmoreceptor present, but perhaps also a Na receptor. Hypertonic saline, hypertonic sucrose or hypertonic glucose, perfused through the cerebral spinal fluid, initiates drinking (Epstein et al., 1969a). However, the most potent of those agents is NaCl (McKinley et al., 1974), an indication that a Na receptor may exist in addition to an osmoreceptor. However, it should be noted that if the receptor senses changes in intracellular Na, any extracellular osmotic stimulus (hypertonic sucrose, glucose, etc.) will increase intracellular Na by pulling water from the cell.

More recently, the effect of various drugs on thirst have been studied by a wide range of authors and laboratories. Perhaps the most potent of all the drugs studied is angiotensin. Epstein et al. (1969) were able to show that an intracranial injection of angiotensin in a rat was an extremely potent stimulus to thirst, with a latency that was as short as only 10 s. If angiotensin was injected intracranially to a starved rat, presented food, the animal stopped eating after only a bite or two, and immediately went to the water bottle in order to drink. This indicates that the angiotensin stimulated thirst so potently that it was able to override what would normally be considered to be an extremely strong drive for food.

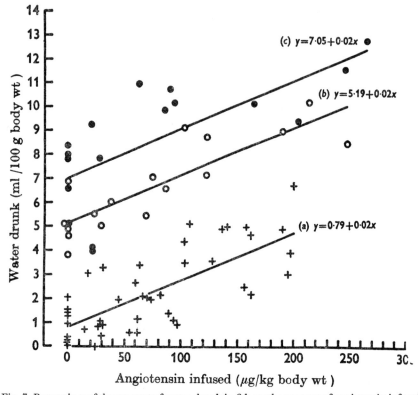

Fig. 7. Regressions of the amount of water drunk in 6 h on the amount of angiotensin infused in nephrectomized rats after intravenous injection of hypertonic NaCl producing increases in osmotic pressure of (a) 0, (b) 5% and (c) 10%. The three regression lines were parallel but significantly separated one from the other, thus supporting the hypothesis of simple additivity of thirst stimuli (from Fitzsimons and Simons, 1969).

These experiments were extended by Fitzsimons and Simons (1969) in experiments in which they injected angiotensin alone or in conjunction with 5% and 10% saline. They were able to show that (Fig. 7) the response to angiotensin infusion is linear, that is, as the dose of angiotensin increased, so did the volume of water drunk. But, furthermore, the curve relating dose of angiotensin to drinking for the three groups of animals, that is, angiotensin alone, angiotensin + 5%

saline or angiotensin + 10% saline, were parallel, i.e. the effects were additive. Brophy (1974), using the cat as a model, was able to show that angiotensin and renin both induced drinking if injected in different locations in the brain. Brophy prepared cats with canulae in different brain areas and injected, through a 30 gauge needle, a small amount of artificial cerebral spinal fluid (CSF) or CSF containing different concentrations of angiotensin or renin. In Fig. 8 are shown the results of his experiments with angiotensin. It may be seen that angiotensin injected in any one of a number of brain areas increased drinking by the cat, and that the drinking was dose related in almost a linear fashion.

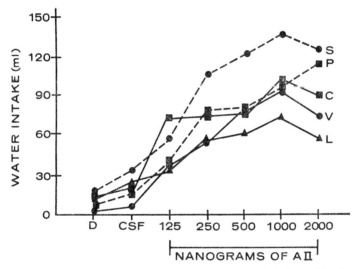

Fig. 8. Angiotensin-induced drinking: volume of water ingested during a 30 min test (D: dummy injection, CSF: artificial cerebrospinal fluid, AII: angiotensin, S: septum, P: preoptic area, C: caudate nucleus, V: lateral cerebral ventricle, L: lateral hypothalamus) (from Brophy and Levitt, 1974).

Similar data were obtained by Brophy (1974) when renin was injected rather than angiotensin. Although the results were similar, it seems unlikely that renin itself is a stimulus to the thirst centre. Rather, it is more likely that the injected renin increased the production rate of brain angiotensin. It has been demonstrated that the brain contains all the enzymes necessary for the conversion of angiotensinogen into angiotensin II (Ganten *et al.*, 1971).

In any event, what does remain clear from these researches is that there appears to be a central thirst receptor which is responsive to the level of angiotensin in the blood or interstitial fluid of the brain such that, when angiotensin levels rise, thirst is stimulated.

More recently it has been demonstrated that endogenously produced angiotensin is a normal stimulator of thirst (Malvin et al., 1977). Rats deprived of water for 30 h were infused through the cerebro-ventricular system with either artificial CSF or CSF plus an angiotensin II blocker, saralasin. Rats infused with the blocker either drank nothing or exhibited attenuated and delayed drinking response. Similar experiments with food-deprived rats had no effect on food intake. Thus it appears that blocking endogenously produced angiotensin significantly decreased drinking, but did not block generalized appetitive behaviour.

Considerable work has also been done with alpha and beta receptor stimulators and it has been demonstrated by many workers that beta adrenergic agonists, such as isoproterenol, are potent stimulators of the thirst centre. If isoproterenol is injected intravenously, the animal is driven to increase the consumption of water (Haupt and Epstein, 1971; Lehr et al., 1967). However, it has also been demonstrated that beta agonists also stimulate the kidney to release renin which in turn increases the circulating level of angiotensin II (Vander, 1965; Ayers, 1969; Ganong, 1972).

Although it appears that renal nerve stimulation can increase renin release, Schrier et al. (1974) suggest that the effect of intravenously administered isoproterenol is not directly on the kidney, but rather indirectly by some undefined extra renal mechanism. This conclusion results from experiments in which isoproterenol was injected either intravenously or directly into the renal artery. At the doses used, only the intravenous route was effective in increasing PRA. However, Vandongen et al. (1973) reported that the isolated perfused rat kidney increases the secretion of renin on beta stimulation, an effect which excludes all extrarenal mechanisms. Thus, many of these experiments in which isoproterenol was injected may be interpreted in a fashion to substantiate the hypothesis that angiotensin is the sole stimulus for drinking. It is quite possible that much of the effect of these drugs is via peripherally generated angiotensin, and it should be noted that, at least in the rat, this appears to be the case. Nephrectomized rats do not increase water consumption when given isoproterenol (Haupt and Epstein, 1971; Meyer et al., 1971). However, rats with their ureters ligated, so that they are functionally anephric do increase water consumption when that drug is given (Haupt and Epstein, 1971). The implication of these experiments is that, in the rat, isoproterenol stimulates the kidney to secrete renin, which in turn increases the circulating level of angiotensin, which in turn stimulates the thirst centre

However, more recent work in the dog seems to imply that the thirst centre is responsive directly to isoproterenol. Fitzsimmons and Sczcepanska-Sadowska (1974) gave isoproterenol intravenously and noted the increase in drinking in these animals. They gave the same dose of drug to nephrectomized dogs. Contrary to what was seen in rats, isoproterinal was quite effective in increasing thirst. Thus, in the dog, isoproterenol appears to have a central action, or at least an action that is independent of angiotensin. In addition, these authors reported that the response to isoproterenol was prevented by the injection of the beta antagonist, propranolol.

In summary, the evidence seems to suggest that the thirst centre is stimulated by a few inputs. A most potent stimulator of thirst is the change in osmotic pressure of fluids bathing the receptor site, a second stimulator of thirst is angiotensin which may be produced by the brain itself or supplied to the brain exogenously, via the kidneys, and finally, beta adrenergic substances are also stimulatory to the thirst centre. All three of these work in concert to regulate the drive for water.

VIII. OTHER HORMONES

A. PROSTAGLANDIN E

Berl and Schrier (1973) infused prostaglandin E_1 (PGE_1) into dogs undergoing a water diuresis. Infusions of 7 μg min^{-1} increased urine osmolality from hypotonic to hypertonic. Similar infusion into hypophysectomized dogs were without effect. These results were interpreted by the authors as evidence that PGE_1 is a control stimulus to ADH release. This interpretation is supported by the experiments of Gagnon et al. (1973). Rat neurohypophysis incubated in vitro with PGE_2 released more ADH into the medium than when incubated in the same medium without added PGE_2. Yamamoto et al. (1978) were able to show that treatment with indomethacin attenuated the stimulatory effect of intraventricular infusions of angiotensin II.

B. ALPHA AND BETA RECEPTORS

The pituitary appears to be affected by both alpha and beta stimulation. Schrier and Berl (1973) found that infusions of noradrenaline into dogs converted hypertonic urine into hypotonic. This change occurred without any change in solute excretion or GFR (an aortic clamp was utilized to maintain renal perfusion pressure constant during the infusion of noradrenaline). When the same dose of noradrenaline was administered to hypophysectomized dogs given ADH, no effect on urine osmolality was noted. One can only conclude from these experiments that noradrenaline inhibits ADH release. Whether this inhibition is at the level of the pituitary or the hypothalamic nuclei is as yet unclear.

Berl et al. (1974) infused either noradrenaline or the beta stimulator, isoproterenol, into normal hydropenic subjects or subjects with diabetes insipidus. In normal subjects, noradrenaline decreased urine osmolality, but diabetes insipidus subjects on ADH showed no response to noradrenaline. Isoproterenol, on the other hand, increased the urine osmolality of normal subjects undergoing water diuresis, but was without effect on the diabetes insipidus subjects. It appears that alpha stimulation is inhibitory while beta stimulation increases the rate of release of ADH.

Beta stimulation is involved in the regulation of ADH secretion in another, indirect way. It has been known for some time that renal nerve stimulation may increase the rate of renin release (Vander, 1965), as does beta-agonist infusion (Ayers, 1969; Gagnon, 1972; Pettinger et al., 1973). Thus, the increased level of angiotensis II may very well serve to stimulate the release of ADH.

C. KALLIKREIN

Infusions of bradykinin or kallidin-10 have been demonstrated to produce a natriuresis (Barraclough and Mills, 1965; Webster and Gilmore, 1964) in the dog. Marin-Grez et al. (1972) and Carretero and Oza (1973) related the urinary excretion of kallikrein to that of Na. If the dogs were saline loaded, Na excretion increased along with kallikrein excretion. Similar findings have been reported by Adetuyibi and Mills (1972) in man. However, Margolis et al. (1974) find an inverse relation in man. It is not clear why there is this discrepancy in results.

Carretero and Aza (1973) believe the kallikrein may be a normal natriuretic factor and propose the scheme in Fig. 9 in which kallikrein exerts its action directly by altering Na transport and indirectly by changing cortical blood flow. Additional support for this scheme comes

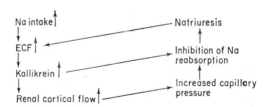

Fig. 9. Kallikrein as a natriuretic factor.

from Marin-Grez (1974). He showed that the ability of rats to excrete a saline load was reduced after they were treated with antibradykinin serum.

D. OXYTOCIN

It has been known for some time that oxytocin was natriuretic and that the natriuresis may result from extrarenal mechanisms (Brooks and Pickford, 1958; Chan and Sawyer, 1961). Brooks and Pickford (1958) reported that small doses of oxytocin injected into the common carotid artery were as effective in producing a natriuresis as much larger doses given intravenously. Furthermore, the intracarotid route did not cause any change in GFR or RPF. Lichardus and Ponec (1972, 1973) gave saline to normal rats, untreated hypophysectomized rats and hypophysectomized rats given a homogenate of pituitary gland intraperitoneally. The first and third groups responded with a natriuresis while the hypophysectomized rats showed no response to saline. It should be noted that only posterior pituitary homogenates were effective. This work is consistent with the hypothesis that oxytocin may be a natriuretic factor, but one which plays a permissive role only. It is difficult to imagine the homogenate liberating variable amounts of a natriuretic factor in response to a saline infusion. Skopkova *et al.* (1973) injected oxytocin into the internal, external or common carotid artery of rats. Only when injected into the common carotid artery was the oxytocin effective in causing natriuresis.

These data support the hypothesis that a chemoreceptor is located in the area of the carotid body which responds to oxytocin. Increasing levels of oxytocin cause the release of another substance or neural output to the kidney such as to reduce Na reabsorption. However, the data are still fragmentary and additional experiments would be necessary to except unequivocably that oxytocin is a physiologically active natriuretic factor.

IX. SUMMARY AND CONCLUSIONS

In Fig. 10 is presented a summary of some of the more important inputs into those systems which regulate body fluid osmolality. The arrows alongside of each parameter represent directional changes, that is increase or decrease in the parameter outlined by the box. The + and −, represent a stimulation or an inhibitory effect on the system to which the arrow points. For example, it may be seen in this diagram that the renin–angiotensin system plays a stimulatory role in the production of ADH. ADH in turn, feeds back upon the renin–angiotensin system and inhibits it. At the same time one can see that the renin–angiotensin system stimulates the thirst centre and also ACTH release. ACTH, in turn, stimulates aldosterone which in turn increases salt appetite, and, if salt appetite is increased, the resulting increase in Na concentration of plasma and in the filtered load of Na inhibits the renin–angiotensin system.

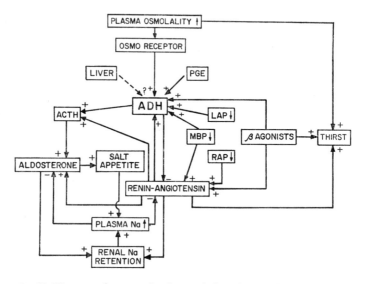

Fig. 10. The control systems for the regulation of salt and water balance.

This scheme shows that a decrease in mean blood pressure stimulates both the release of ADH and of renin. A reduction in right atrial pressure stimulates the release of renin. The dashed line connecting the box labelled liver to ADH represents a possible efferent input from the liver to the hypothalamic posterior pituitary axis. The dashed line indicates that delineation of that system remains to be worked out more clearly.

It should be emphasized that this scheme is highly simplified. Undoubtedly, there are many more interrelations between the hormones and the receptor systems described in this figure in addition to many others deliberately omitted, and those that remain to be discovered. In any event, I feel that this representation presents most of the important control systems regulating those hormones which have as their major function the control of the body's salt and water balance.

X. MARINE CARNIVORES

Although mammals which live in the sea have essentially the identical problems in osmoregulation as those of terrestrial mammals, the major difference between them lies in the unique environment of the former. Marine mammals lack one important area of osmoregulatory stress to which all terrestrial mammals are exposed. Animals of the sea face essentially no difficulty in maintaining body temperature constant without the imposition of water loss, either through sweat or panting. Thus, regardless of their environmental temperature, their own body temperature is maintained constant by the conservation of heat rather than by regulating heat loss through evaporation of sweat or evaporation through the respiratory tract. In addition, marine mammals lose considerably less water than terrestrial mammals during respiration This is for two reasons, the first being that the environment of the marine mammals is generally heavily laden with water, that is the vapour pressure of the inspired air is generally considerably higher than that which most terrestrial mammals inhale. Secondly, marine mammals have an unusual trick of respiration which serves to reduce respiratory water loss. Respiration in animals such as the porpoise, whale and seal is not only irregular, but quite slow. Resting respiratory rates in cetacia are on the order of one or two breaths min^{-1} (Ridgway, 1972). An animal such as the porpoise has a large tidal volume which it holds in its lungs for a relatively long time so that a much larger fraction of the inspired oxygen is removed in the lungs than is so for terrestrial mammals. What this means is that for any given oxygen consumption, the volume of air moved in and out of the lungs is less for a marine mammal than for a terrestrial one. In this fashion, a considerable amount of respiratory water loss is saved by the marine mammal. Compared to terrestrial mammals evaporative loss is reduced as much as 77% (Coulcombe et al., 1965). What all this adds up to is that animals living in the sea have, in general, a lower requirement for the supply of fresh water than do terrestrial mammals. This is not to say that the marine environment imposes no problems of water balance on mammals living in the sea, rather that their problems are of a different order of magnitude and can be solved in a different way.

A series of investigators have made extensive studies of the water balance of marine mammals by measuring the urinary losses plus loss of water via the faeces and through the respiratory tract. By calculating the amount of water contained in the fish eaten by marine mammals each day, and adding to that the metabolic water derived from the metabolism of the fat and carbohydrate sources in those fish, one obtains a figure for water which is equal to that, and sometimes greater, than the calculated water loss. Thus, it is evident that marine mammals are able to obtain sufficient water from their food supplies to maintain a state of constant osmolality of body fluids (Irving et al., 1935). Perhaps this calculation is a bit superfluous; it is perfectly obvious that marine mammals do live and survive well in the marine environment without a source of fresh water. The calculations only confirm the obvious. Marine mammals are capable of surviving without recourse to a supply of fresh water. The more interesting problems concerning osmoregulation in marine mammals is, however, that of trying to describe those regulatory systems which serve to change the output of urine in response to changing loads of water and salt imposed upon the animal following feeding. Although very considerable amounts of information are lacking, there does appear in the literature a sufficient amount of work to more than whet the appetite of those interested in osmoregulation of marine mammals. Some of those experiments are described below.

A. RENAL FUNCTION

It has been known for years (Smith, 1936; Hiatt and Hiatt, 1942) that the rate of urine production in marine mammals increases very significantly following a meal of fish, and then slowly decreases to control values. Coincident with this diuresis there is an increase in glomerular filtration rate and renal plasma flow. These early investigators assumed that the increase in urine flow following a meal was a result of the simultaneous increase in glomerular filtration rate and renal plasma flow. However, it was shown by Malvin and Rayner (1968) that although GFR and RPF do increase along with urine flow following a meal, renal function returns to control values within 4 to 5 h, while urine flow and Na excretion remain elevated for a much longer period of time. In addition, if one plotted the time course of increase and decrease of urine flow, Na excretion, GFR and RPF, there was not a one-to-one relationship between the haemodynamic changes and excretory rates, even in the early phases following a meal. Thus, although some of the diuresis can be ascribed to a doubling of the GFR early in time, the major changes in diuresis and natriuresis cannot be the result of altered GFR and RPF. Something else must be involved.

Fetcher and Fetcher (1942) demonstrated that porpoises will exhibit a diuresis and natriuresis if given 0·5 M NaCl orally. Similar studies have been performed by Ridgway (1972) in different species of cetacia. He was able to show that if an animal is given sea water by stomach tube, similar changes in urine flow are obtained. Additionally, Na excretion also increases following the instillation of sea water into the stomach. There is noted, however, in these experiments, a paradoxical situation. In marine mammals given fish, urine osmolality remained essentially constant even though urine flow rates increased by a factor of 10, a finding also reported by Malvin and Rayner (1968). Animals forced to consume sea water showed a different pattern of response. In these animals there was a negative correlation of U_{osm} with urine flow. Two typical experiments of this sort are illustrated in Fig. 11.

It can be seen that following a test meal of 5 kg of fish, there was no

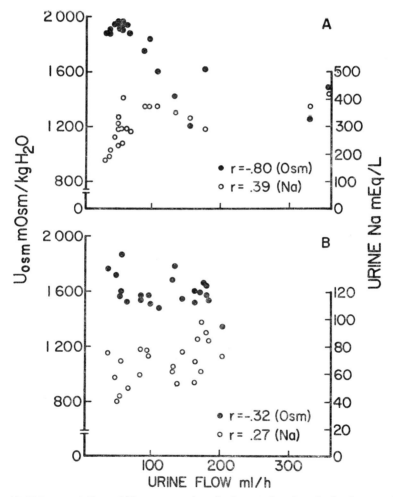

Fig. 11. Urine osmolality and Na concentration of urine as a function of urine flow rate in a 77 kg female *Tursiops truncatus* fed 5 kg of Spanish mackerel or 3 l of sea water by stomach tube, (A) seawater; (B) mackerel (plotted from data in Ridgway, 1972).

significant correlation between the rate of urine flow and urine osmolality or Na concentration in the urine over the 24 h period followed in these experiments. This is consistent with the old observation of Smith (1936) that in the seal no relation was found between urine flow rate and the depression of the freezing point of urine. On the other hand, the same animal given 3 l of sea water by stomach tube exhibited a significant inverse relationship between urine osmolality and urine flow with no significant correlation between urine flow and Na concentration. Even in this animal, in which there was a significant correlation coefficient for urine osmolality and urine flow rate, the decrease in osmolality was minimal. Urine flow rate changed by a factor of 10, yet urine osmolality fell from about 1900 mosmol kg^{-1} H$_2$O to a minimum value of only about 1200 mosmol kg^{-1} H$_2$O. It would appear that the response to sea water is somewhat different from the response to feeding. The reason for this remains unclear.

A third type of experiment done with porpoises was to follow the rate of urine flow, urine osmolality and Na excretion upon ingestion of fresh water. Malvin and Rayner (1968) reported that 4 l of fresh water placed into the stomach of a porpoise resulted in no detectable change in urine flow rate or urine osmolality. However, Ridgway (1972) performed similar experiments on a series of porpoises and in all cases reported that urine flow does increase when fresh water is placed in the stomach of a porpoise. Furthermore, he was able to show that urine osmolality fell as urine flow rate increased. It is noteworthy, however, that even though the animals appeared to undergo a rather severe water diuresis (urine flow rates increased over five-fold) urine osmolality never became hypotonic. In the series of experiments reported, the lowest urine osmolality achieved was 482 mosmol kg^{-1} H$_2$O, a value significantly above plasma osmolality.

B. ANTIDIURETIC HORMONE

Although the porpoise is able to change urine flow rates very considerably in response to ingestion of food or water, that response does not appear to be dependent upon ADH levels in the blood, and in fact there is some question as to whether or not cetacia have any circulating ADH. Geiling et al. (1940) stated, without giving data, that the ADH concentrations in the posterior lobe of porpoise pituitaries were extremely low. Malvin et al. (1971) analysed the pituitaries of two porpoises and reported that they contained relatively little antidiuretic material. On a weight basis the concentration was about $\frac{1}{20}$ of that in rat pituitaries. Since the estimation depends upon a bioassay, it is not clear that this was indeed antidiuretic hormone. The apparent antidiuretic response could have been due to hormones other than ADH, (oxytocin). More to the point, these same authors measured the concentration of bioassayable ADH in the plasma of six different animals following a fast lasting up to 60 h. In the 15 blood samples analysed, detectable antidiuretic material was obtained in only three samples, and in those samples the level was only 0·2 and 0·3 μU ml^{-1}. These levels are so low as to make one wonder whether it was indeed ADH or some contaminant. In any event, it appears that if cetacia do respond to ADH they must be exquisitely sensitive to that hormone, as the levels are far lower in cetacia than in any animal studied to date. In this context it is worth quoting Ridgway (1972):

> In my view, the absence of ADH in significant quantities is not an incongruous discovery, particularly when the overall physiology of the aquatic mammal is considered. In his natural environment of sea water, the marine cetacia is suspended in a state very close to neutral buoyancy. Therefore, his body as a whole is in a state of weightlessness. In man one of the most striking physiological manifestations of immersion is diuresis. Immersion is thought to cause a hydrostatic pressure gradient forcing blood from the veins of the lower extremities into the central circulation. The consequent distention of cardiac atrial stretch receptors by a concurrent increase in intrathoracic blood volume causes an inhibition of the release of ADH with the resulting diuresis. A similar problem of diuresis results from weightlessness of space flight. ADH depression may be a natural consequence of immersion in water, but porpoises apparently replace ADH with some other mechanism for the control of diuresis.

It is interesting to note that the seal in contrast to cetacia, may have a physiologically active ADH system. Bradley *et al.* (1954) reported that the diuresis established in a seal by water loading was inhibited by pitressin. However, it should also be noted that the seal is not confined to the sea, but does spend a fraction of its life as a terrestrial mammal.

C. RENIN–ALDOSTERONE

Although it is evident that marine mammals respond rapidly to ingestion of Na and water by a natriuresis and diuresis the physiological mechanisms involved in those responses are not known. It might be inferred that the renin–angiotensin–aldosterone system is, at least in part, responsible for the natriuresis and diuresis. Unfortunately, minimal data are available. Malvin and Vander (1967b) reported that in three species of porpoises and two whales the mean postabsorptive PRA was 2·6 ± 1·4 ng ml^{-1} h^{-1}. This is a value well within the normal range for terrestrial mammals. Ridgway (1972) reports that urinary levels of aldosterone are also in the same range one would expect for land carnivores. Finally, data of Malvin *et al.* (1978) indicate that PRA and plasma aldosterone levels in cetacia change in the same direction during fasting and after water loading. These data are support for the hypothesis that in marine mammals the renin–aldosterone link is present. Whether or not this is the sole means by which these animals regulate salt and water balance is as yet unclear. It could very well be that some other as yet undefined mechanism must also be present.

The urinary response to sea water and eating are different in that following eating no relation exists between U_{osm} and urine flow, while after sea water an inverse relation exists. This is suggestive of a different mechanism being involved in the responses. In addition, the increase in urine flow and Na excretion is so great and so rapid following feeding that it appears unlikely that a reduction in aldosterone levels of plasma by itself could explain the response. Within 4 h Na excretion rises from about 1% of the filtered load to approximately 6% (Malvin and Rayner, 1968), an increase which seems too high to be solely the result of a reduction in circulating aldosterone levels. Also, K excretion does not show an inverse relation to Na excretion as would be expected if the response were due only to reduced aldosterone levels. It seems then, that cetacia regulate salt and water balance in a way which incorporates the renin–aldosterone system, but which also uses another undefined control system, perhaps even the elusive natriuretic hormone.

REFERENCES

Adetuyibi, A. and Mills, I. H. (1972). Relation between urinary kallikrein and renal function, hypertension and excretion of sodium and water in man. *Lancet* **2**, 203–207.

Akachi, A., Niijima, A. and Jacobs, H. L. (1976). An hepatic osmoreceptor mechanism in the rat: electrophysiological and behavioural studies. *Am. J. Physiol.* **231**, 1043–1049.

Andrews, W. H. H. and Orbach, J. (1974). Sodium receptors activating some nerves of perfused rabbit livers. *Am. J. Physiol.* **227**, 1273–1275.

Annat, G., Grandjean, B., Vincent, M., Jarsaillon, E. and Sassard, J. (1976). Effects of right atrial stretch on plasma renin activity. *Arch. Int. Physiol. Biochem.* **84**, 311–315.

Ayers, C. R., Harris, R. H. and Lefer, L. G. (1969). Control of renin release in experimental hypertension. *Circ. Res.* **24**, 103–112.

Barraclough, M. A. and Mills, I. H. (1965). Effect of bradykinin on renal function. *Clin. Sci.* **28**, 69–74.

Barraclough, M. A., Jones, N. F. and Marsden, C. D. (1967). Effect of angiotensin on renal function in the rat. *Am. J. Physiol.* **212**, 1153–1157.

Berl, T. and Schrier, R. W. (1973). Mechanism of effect of prostaglandin E_1 on renal water excretion. *J. clin. Invest.* **52**, 463–471.

Berl, T., Hardbottle, J. A. and Schrier, R. W. (1974). Effect of alpha and beta-adrenergic stimulation on renal water excretion in man. *Kidney Int.* **6**, 247–253.

Bonjour, J. P. and Malvin, R. L. (1970). Stimulation of ADH release by the renin angiotensin system. *Am. J. Physiol.* **218,** 1555–1559.

Bradley, S. E., Mudge, G. H. and Blake, W. D. (1964). The renal excretion of sodium, potassium and water by the harbour seal (Phoca vit ulina L.): effect of apnea, sodium, potassium and water loading; pitressin and mercurial diuresis. *J. Cell comp. Physiol.* **43,** 1–22.

Brennan, L. A., Malvin, R. L., Jochim, K. E. and Roberts, D. E. (1971). Influences of left and right atrial receptors on plasma concentrations of ADH and renin. *Am. J. Physiol.* **221,** 273–278.

Brooks, C. McC., Ishikawa, T., Koizumi, K. and Lu, H. H. (1966). Activity of neurones in the paraventricular nucleus of the hypothalamus and its control. *J. Physiol. (Lond.)* **182,** 217–231.

Brooks, F. P. and Pickford, M. (1958). The effect of posterior pituitary hormones on the excretion of electrolytes in dogs. *J. Physiol. (Lond.)* **142,** 468–493.

Brophy, P. D. and Levitt, R. A. (1974). Dose-response analysis of angiotensin- and renin-induced drinking in the cat. *Pharmacol. Biochem. Behav.* **2,** 509–514.

Bunag, D. D., Page, T. H. and McCubbin, J. W. (1967). Inhibition of renin release by vasopressin and angiotensin. *Cardiovascular Res.* **1,** 67–73.

Carretero, O. A. and Aza, N. B. (1973). Urinary kallikrein, sodium metabolism and hypertension. *Proc. Int. Congr. Series* **302,** Los Angeles, 290–299.

Chan, W. Y. and Sawyer, W. H. (1961). Saluretic actions of neurohypophysial peptides in conscious dogs. *Am. J. Physiol.* **201,** 799–803.

Clark, B. J. and Silva, M. R. (1967). An efferent pathway for the selective release of vasopressin in response to carotid occlusion and hemorrhage in the cat. *J. Physiol. (Lond.)* **191,** 529–542.

Claybaugh, J. (1976). Effect of dehydration on stimulation of ADH release by heterologous renin infusions in conscious dogs. *Am. J. Physiol.* **231,** 655–660.

Claybaugh, J. R., Share, L. and Shimizu, K. (1972). The inability of infusions of angiotensin to elevate the plasma vasopressin concentration. *Endocrinology* **90** 1647–1652.

Coulombe, H. N., Ridgway, S. H. and Evans, W. E. (1965). Respiratory water exchange in two species of porpoise. *Science N.Y.* **149,** 86–88.

Cross, B. A. and Green, J. D. (1959). Activity of single neurones in the hypothalamus: effect of osmotic and other stimuli. *J. Physiol. (Lond.)* **148,** 554–569.

Daly, J. J., Roe, J. and Horrocks, P. (1967). A comparison of sodium excretion following the infusion into systemic and portal veins in the dog. *Clin. Sci.* **33,** 481–487.

Daniels-Severs, A., Ogden, E. and Vernikos-Daniells, J. (1971). Centrally mediated effects of angiotensin II in the unanesthetized rat. *Physiol. Behav.* **7,** 785–787.

Dennhardt, R., Ohm, W. W. and Haberich, F. J. (1971). Die ausschaltung ter leberäste des N. vagus an der wachen ratte und ihr einfluss auf die hepatogene diurese—indirekter beweis fur die afferente leitung der leberosmoreceptoren uber den N. vagus. *Pflugers Arch.* **328,** 51–56.

Dickinson, C. J. and Lawrence, J. R. (1963). A slowly developing pressor response to small concentrations of angiotensin—its bearing on the pathogenesis of chronic renal hypertension. *Lancet* **1,** 1354–1365.

Dunn, F. L., Brennan, T. J., Nelson, A. E. and Robertson, G. L. (1973). The role of blood osmolality and volume in regulating vasopressin secretion in the rat. *J. clin. Invest.* **52,** 3212–3219.

Emmers, R. (1973). Interaction of neural systems which control body water. *Brain Res.* **49,** (2), 323–347.

Epstein, A. N., Fitzsimons, J. T. and Simons, B. T. (1969b). Drinking caused by the intracranial injection of angiotensin into the rat. *J. Physiol. (Lond.)* **200**, 98–100.

Fetcher, E. S. and Fetcher, G. W. (1942). Experiments on the osmotic regulation of dolphins. *J. Cell comp. Physiol.* **19**, 123–130.

Fitzsimons, J. T. (1961). Drinking by nephrectomized rats injected with various substances. *J. Physiol. (Lond.)* **155**, 563–569.

Fitzsimons, J. T. (1963). The effects of slow infusions of hypertonic solutions on drinking and drinking thresholds in rats. *J. Physiol. (Lond.)* **167**, 344–354.

Fitzsimons, J. T. (1971). The hormonal control of water and sodium intake. In "Frontiers in Neuroendocrinology" (L. Martini and W. F. Ganong, eds). Oxford University Press, New York and Oxford, pp. 103–128.

Fitzsimons, J. T. and Simons, B. J. (1969). The effect on drinking in the rat of intravenous infusions of angiotensin given alone or in combination with other stimuli of thirst. *J. Physiol. (Lond.)* **203**, 45–57.

Fitzsimons, J. T. and Szczepanska-Sadowska, E. (1974). Drinking and antidiuresis elicited by isoprenaline in the dog. *J. Physiol. (Lond.)* **239**, 251–267.

Gagnon, D. J., Cousineau, D. and Brecher, P. J. (1973). Release of vasopressin by angiotensin II and prostaglandin E_2 from the rat hypophysis in vitro. *Life Sci.* **12**, 487–497.

Ganong, W. F. (1972). Sympathetic effects on renin secretion: mechanism and physiological role. In "Control of Renin Secretion" (T. A. Assaykeen, ed.). Plenum Press, New York.

Ganten, G. J., Minnich, J. L., Granger, P., Hayduk, K., Brecht, H. M., Barbeau, A., Boucher, R. and Genest, J. (1971). Angiotensin-forming enzyme in brain tissue. *Science N.Y.* **173**, 64–65.

Gauer, O. H., Henry, J. P., Sieker, H. O. and Wendt, W. F. (1954). The effect of negative pressure breathing on urine flow. *J. clin. Invest.* **33**, 287.

Gillespie, D. J., Sandberg, R. L. and Koike, T. I. (1973). Dual effect of left atrial receptors on excretion of sodium and water in the dog. *Am. J. Physiol.* **225**, 706–710.

Glasby, M. A. and Ramsay, D. J. (1974). Hepatic osmoreceptors? *J. Physiol. (Lond.)* **243**, 765–776.

Goetz, K. L., Hermreck, A. S., Slick, G. L. and Starke, H. S. (1970). Atrial receptors and renal function in conscious dogs. *Am. J. Physiol.* **219**, 1417–1423.

Gregg, C. M. and Malvin, R. L. (1974). Specific binding of angiotensin II to neurohypophysis not correlated with stimulation of ADH release. *Physiologist* **17**, 232.

Gregg, C. M. and Malvin, R. L. (1978). Localization of central sites of action of angiotensin II on ADH release in vitro. *Am. J. Physiol.* **234**, F135–F140.

Grier, M. A. (1955). Suggestive evidence of a primary "drinking center" in hypothalamus of the rat. *Proc. Soc. Exp. Biol. Med.* **89**, 59–62.

Haberich, F. J. (1968). Osmoreception in the portal circulation. *Fed. Proc.* **27**, 1137–1141.

Haberich, F. J. (1971). Osmoreceptors in the portal circulation and their significance for the regulation of water balance. *Triangle* **10**, 123–130.

Haberich, F. J., Aziz, O. and Nowacki, P. E. (1964). Über einen osmoreceptosrich tatigen mechanismus in der leber. *Arch. Grs. Physiol.* **285**, 73–89.

Hayward, J. N. and Vincent, J. D. Osmosensitive single neurones in the hypothalamus of unanesthetized monkeys. *J. Physiol., Lond.* **210**, 947–972.

Henry, J. P. and Pearce, J. W. (1956). The possible role of atrial stretch receptors in the induction of changes in urine flow. *J. Physiol. (Lond.)* **131**, 572–585.

Hiatt, E. P. and Hiatt, R. B. The effect of food on the glomerular filtration rate and renal blood flow in the harbour seal. *J. Cell comp. Physiol.* **19**, 221–227.

Holmes, J. H. and Gregersen, M.I. (1950a). Observations on drinking induced by hypertonic solutions. *Am. J. Physiol.* ECT, **162** 326–337.

Holmes, J. H. and Gregersen, M. I. (1950b). Role of sodium and chloride in thirst. *Am. J. Physiol.* **162**, 338–347.

Houpt, K. A. and Epstein, A. N. (1971). The complete dependence of beta-adrenergic drinking on the renal dipsogen. *Physiol. Behav.* **7**, 897–902.

Irving, L., Fisher, K. C. and McIntosh, F. C. (1935). Water balance of a marine mammal, the seal. *J. CCP* **6**, 387–391.

Johnson, J. A., Moore, W. W. and Segar, W. E. (1969). Small changes in left atrial pressure and plasma antidiuretic hormone titers in the dog. *Am. J. Physiol.* **217**, 210–214.

Johnson, M. D. and Malvin, R. L. (1975). The antidiuretic and antinatriuretic effect of intrarenal arterial infusion of angiotensin II. *Fed. Proc. (abstract).* **34**, 377.

Kappagoda, C. T., Linden, R. J. and Snow, H. M. (1973). Effect of stimulating right atrial receptors on urine flow in the dog. *J. Physiol. (Lond.)* **235**, 493–502.

Kappagoda, C. T., Linden, R. J., Snow, H. M. and Whitaker, E. M. (1975). Effect of destruction of the posterior pituitary on the diuresis from left atrial receptors. *J. Physiol. (Lond.)* **244**, 757–770.

Keil, L. C., Summy-Long, J. and Severs, W. B. (1975). Release of vasopressin by angiotensin II. *Endocrinology* **96**, 1063–1065.

Kinney, M. J. and DiScala, V. A. (1972). Renal clearance studies of the effect of left atrial distention in the dog. *Am. J. Physiol.* **222**, 1000–1003.

Koizumi, K., Ishikawa, T. and Brooks, C. McC. (1964). Control of neurons in the supraoptic nucleus. *J. Neurophysiol.* **27**, 878–892.

Laragh, J. H., Angers, M., Kelly, W. G. and Lieberman, S. (1960). Hypotensive agents and pressor substances: Effect of epinephrine, norepinephrine, angiotensin II and others on the secretory rate of aldosterone. *JAMA* **174**, 234–240.

Ledsome, J. R. and Mason, J. M. (1972). The effect of vasopressin on the diuretic response to left atrial distension. *J. Physiol. (Lond.)* **221**, 427–440.

Lehr, D., Mallow, J. and Krukowski, M. (1967). Copious drinking and simultaneous inhibition of urine flow elicited by beta-adrenergic stimulation and contrary effect of alpha-adrenergic stimulation. *J. Pharmacol. Exp. Ther.* **158**, 150–163.

Lichardus, B. and Ponec, J. (1972). Neurohypophyseal origin of a humoral factor restoring volume natriuresis in acutely hypophysectomized rats. *Experientia* **28**, 1443–1444.

Lichardus, B. and Ponec, J. (1973). On the role of the hypophysis in the renal mechanism of body fluid volumes regulation. *Endokrinologie* **61**, 403–412.

Lydtin, H. (1969a). Untersuchugen über mechanismen der osmo-und volumenregulation II. Untersuchen über der einfluss intravenös intraportal und oral zugeführter hypotoner Kochsalzlosungen auf die diurese des Hurdes. *Z. ges. exp. Med.* **149**, 193–210.

Lydtin, H. (1969b). Untersuchugen über mechanismen der osmo-und volumenregulation 3. Untersuchungen am mechen über die wirkung von oral und intravenös jugefuhrter Kochsalzlösungen auf die Harvausscheidung. *Z. ges. exp. Med.* **149**, 211–225.

Lydtin, H. and Hamilton, W. F. (1964). Effect of acute changes in left atrial pressure on urine flow on unanesthetized dogs. *Am. J. Physiol.* **207**, 530–536.

McKinley, M. J., Blaine, E. H. and Denton, D. A. (1974). Brain osmoreceptors, cerebrospinal fluid electrolyte composition and thirst. *Brain Res* **70**, 532–537.

Malvin, R. L. (1972). Angiotensin and antidiuretic hormone. In "Endocrinology" (R. O. Scow, ed.). Amsterdam Press, Amsterdam, pp. 717–723.
Malvin, R. L. and Rayner, M. (1968). Renal function and blood chemistry in cetacea. Am. J. Physiol. 214, 187–191.
Malvin, R. L. and Vander, A. J. (1967a). Effects of angiotensin infusion on renal function in the unanesthetized rat. Am. J. Physiol. 213, 1205–1208.
Malvin, R. L. and Vander, A. J. (1967b). Plasma renin activity in marine teleosts and cetacea. Am. J. Physiol. 213, 1582–1584.
Malvin, R. L., Bonjour, J.-P. and Ridgway, S. H. (1971). Antidiuretic hormone levels in some cetaceans. Proc. Soc. exp. Biol. Med. 136, 1203–1205.
Malvin, R. L., Mouw, D. R. and Vander, A. J. (1977). Angiotensin: Physiological role in water deprivation induced thirst in rats. Science. N.Y. 197, 171-173.
Malvin, R. L., Ridgway, S. H. and Cornell, L. (1978). Renin and aldosterone levels in dolphins and sea lions. Proc. Soc. exp. Biol. Med. 157, 665–668.
Margolius, H. S., Horwitz, D., Geller, R. G., Alexander, R. W., Gill, J. R., Pisano, J. J. and Keiser, H. R. (1974). Urinary kallikrein excretion in normal man. Circ. Res. 35, 812–819.
Marin-Grez, M., Cottone, P. and Carretero, O. A. (1972). Evidence for an involvement of kinins in regulation of sodium excretion. Am. J. Physiol. 223, 794–796.
Marin-Grez, M. (1974). The influence of antibodies against bradykinin on isotonic saline diuresis in the rat. Pflügers Arch. 350, 231–239.
Meyer, D. K., Peskar, B. and Hertting, G. (1971). Hemmung des durch blutdrucksenkende Pharmak bei ratten ausgelösten trinkens durch nephrecktomie. Experientia 7, 65–66.
Mouw, D., Bonjour, J. P., Malvin, R. L. and Vander, A. (1971). Central action of angiotensin in stimulating ADH release. Am. J. Physiol. 220, 239–242.
Mulrow, P. J. and Ganong, W. F. (1961). The effect of hemorrhage upon aldosterone secretion in normal and hypophysectomized dogs. J. clin. Invest. 40, 579–585.
Mulcahy, J. J., Geis, W. P. and Malvin, R. L. (1975). The effect of cardiac denervation on renal function. Proc. Soc. exp. Biol. Med. 149. In press.
Nicoll, R. A. and Barker, J. L. (1971). Excitation of supraoptic neurosecretory cells by angiotensin II. Nature, Lond. 233, 172.
Nonidez, J. F. (1937). Identification of the receptor areas in the vena cavae and pulmonary veins which initiate reflex cardiac acceleration (Bainbridge's Reflex). Am. J. Physiol. 61, 203–231.
Oelkers, W., Brown, J. J., Fraser, R., Lever, A. F., Morton, J. J. and Robertson, J. I. S. (1974). Sensitization of adrenal cortex to angiotensin II in sodium-depleted man. Circ. Res. 34, 69–77.
Olsson, K. and Kolmodin, R. (1974). Accentuation by angiotensin II of the antidiuretic and dipsogenic responses to intracarotid infusions of NaCl and fructose. ACTA Endocrin. 75, 333–341.
Passo, S. S., Thornborough, J. R. and Rothballer, A. B. (1972). Hepatic receptors in control of sodium excretion in anesthetized cats. Am. J. Physiol. 224, 373–375.
Pettinger, W. A., Campbell, W. B. and Keeton, K. (1973). Adrenergic component of renin release induced by vasodilating antihypertensive drugs in the rat. Circ. Res. 33, 82–86.
Potkay, S. and Gilmore, J. P. (1970). Renal responses to vena caval and portal venous infusions of sodium chloride in unanesthetized dogs. Clin. Sci. 39, 13–20.
Ridgway, S. H. (1972). "Mammals of The Sea" (S. H. Ridgway, ed.) Charles C. Thomas, Springfield, Illinois.

Sakai, K. K., Marks, B. H., George, J. and Koestner, A. (1974). Specific angiotensin II receptors in organ-cultured canine supraoptic nucleus cells. *Life Sci.* **14**, 1337–1344.

Schneider, E. G., Davis, J. O., Robb, C. A., Baumer, J. S., Johnson, J. A. and Wright, F. S. (1970). Lack of evidence for a hepatic osmoreceptor mechanism in conscious dogs. *Am. J. Physiol.* **218**, 42–45.

Schrier, R. W. and Berl, T. (1973). Mechanism of effect of alpha adrenergic stimulation with norepinephrine on renal water excretion. *J. clin. Invest.* **52**, 502–511.

Schrier, R. W., Hardbottle, J. and Berl, T. (1974). Influence of the adrenergic nervous system on renal water excretion and renal renin secretion. In "Recent Advances in Renal Physiology and Pharmacology" (L. G. Wesson and G. M. Fanelli, eds). University Park Press, Baltimore.

Severs, W. B., Daniels, A. E. and Buckley, J. P. (1967). On the central hypertensive effect of angiotensin II. *Int. J. Neuropharmac.* **6**, 199–205.

Severs, W. B., Summy-Long, J., Taylor, J. S. and Connor, J. D. (1970). A central effect of angiotensin: release of pituitary pressor material. *J. Pharm. exp. Ther.* **174**, 27–34.

Severs, W. B., Daniels-Severs, A., Summy-Long, J. and Radio, G. J. (1971). Effects of centrally administered angiotensin II on salt and water excretion. *Pharmacology* **6**, 242–252.

Share, L. (1965). Effect of carotid occlusion and left atrial distension on plasma vasopressin titer. *Am. J. Physiol.* **208**, 219–223.

Share, L. (1967). Role of peripheral receptors in the increased release of vasopressin in response to hemorrhage. *Endocrinology* **81**, 1140–1146.

Share, L. (1968). Control of plasma antidiuretic hormone titer in hemorrhage: role of atrial and arterial receptors. *Am. J. Physiol.* **215**, 1384–1389.

Share, L. and Levy, M. N. (1962). Cardiovascular receptors and blood titer of antidiuretic hormone. *Am. J. Physiol.* **203**, 425–428.

Share, L. and Levy, M. N. (1966). Carotid sinus pulse pressure, a determinant of plasma antidiuretic hormone concentration. *Am. J. Physiol.* **211**, 721–724.

Shimizu, K., Share, L. and Claybaugh, J. R. (1973). Potentiation by angiotensin II of the vasopressin response to an increasing plasma osmolality. *Endocrinology* **93**, 42–50.

Skopkova, J., Albrecht, I. and Cort, J. H. (1973). A natriuresis receptor at the carotid bifurcation specifically activated by oxytocin. *Pflugers Arch.* **343**, 123–132.

Smith, H. W. (1936). The composition of urine in the seal. *J. cell comp. Physiol.* **7**, 465–473.

Smookler, H. H., Severs, W. B., Kinnard, W. J. and Buckley, J. P. (1966). Centrally mediated cardiovascular effects of angiotensin II. *J. Pharmac. exp. Ther.* **153**, 485–494.

Stitzer, S. O. and Malvin, R. L. (1975). Right atrium and renal sodium excretion. *Am. J. Physiol.* **222**, 184–190.

Strandhoy, J. W. and Williamson, H. E. (1970). Evidence for an hepatic role in the control of sodium excretion. *Proc. Soc. exp. Biol. Med.* **133**, 419–422.

Tagawa, H., Vander, A. J., Bonjour, J-P. and Malvin, R. L. (1971). Inhibition of renin secretion by vasopressin in unanesthetized sodium-deprived dogs. *Am. J. Physiol.* **220**, 949–951.

Uhlich, E., Weber, P. and Gröschel-Stewart, U. (1974). Angiotensin-stimulated vasopressin release in man; radioimmunologically determined plasma levels of vasopressin. *Acta endocr.* **184**, 52.

Vander, A. J. (1965). Effect of catecholamines and the renal nerves on renin secretion in the anesthetized dog. *Am. J. Physiol.* **209**, 659–662.

G

Vander, A. J. (1968). Inhibition of renin release in the dog by vasopressin and vasotocin. *Circ. Res.* **23**, 605–609.

Vandongen, R., Peart, W. S. and Boyd, G. W. (1973). Adrenergic stimulation of renin secretion in the isolated perfused rat kidney. *Circ. Res.* **32**, 290–296.

Verney, E. B. (1947). The antidiuretic hormone and the factors which determine its release. *Proc. Royal Soc., Lond. Ser. B.* **135**, 25.

Walters, J. K. and Hatton, G. I. (1974). Supraoptic neuronal activity in rats during five days of water deprivation. *Physiol. Behav.* **13**, 661–667.

Webster, M. E. and Gilmore, J. P. (1964). Influence of kallidin-10 on renal function. *Am. J. Physiol.* **206**, 714–718.

Yamamoto, M., Share, L. and Shade, R. E. (1978). Effect of ventriculo-cisternal perfusion with angiotensin II and indomethacin on the plasma vasopressin concentration. *Neuroendocrinology* **25**, 166–173.

Zehr, J. E., Johnson, J. A. and Moore, W. W. (1969). Left atrial pressure, plasma osmolality, and ADH levels in the unanesthetized ewe. *Am. J. Physiol.* **217**, 1672–1680.

5. Mammalian Herbivores

G. M. O. MALOIY, W. V. MACFARLANE and A. SHKOLNIK

University of Nairobi, Nairobi, Kenya, Waite Agricultural Research Institute, Adelaide, South Australia and Tel-Aviv University, Tel-Aviv, Israel

I. Introduction	185
II. Osmoregulation of Mammalian Herbivores	188
A. Diffusion	189
B. Secretion	191
C. Cell Osmoregulation	193
References	207

I. INTRODUCTION

The study of mammalian osmoregulation has several aspects. In part it is concerned with maintenance of cellular and extracellular osmotic pressures within biologically viable limits. In part it is also a study of the adaptive and evolutionary aspects of rate functions in a range of diverse animals. Another osmoregulatory component is the input–output budget of water and electrolytes, as well as their adjustment to environmental extremes. Attention will be given to these topics in this chapter.

Mammalian herbivores comprise representatives of animal groups differing not only in body size but also in functions and survival mechanisms. This chapter aims to discuss osmoregulation in such herbivores as hyrax, equids, camelids, as well as wild and domestic ruminants.

Herbivores have radiated into a variety of geographical and ecological niches. Their range of distribution encompasses the Arctic tundra or taiga, cool, temperate forests, grasslands, savannah, wet tropical lands and hot, dry deserts. In these regions, herbivores are exposed not only to climatic extremes but also to varying quality and quantity of forage and nutrients. Algae, lichens, mosses, leaves, stems, vines, succulents, forbs, grasses, desert scrub and saltbush with varying water and mineral contents, are all exploited by herbivorous mammals as sources of nutrients.

The position of herbivores in the trophic hierarchy as intermediaries in the transfer of energy from the plant producers to the consumers of animals, lends this group its significance in every ecosystem and in the economy and welfare of almost all human civilizations.

Arctic cold as well as the heat of deserts, impose special physiological strains on these mammals, particularly in regulation of temperature, water and salt balance. Water is quantitatively the main constituent of the animal body. In lean herbivores, however, water as a percentage of body weight is generally greater than in most other mammals (Table I) because of the presence of large specialized fermentation zones containing water (rumen, caecum, colon) in the alimentary tract.

Although herbivores subsist on plants, some select herbage with a high water content while others live on xerophytes. In humid areas, for example, herbivores which feed on the leaves of succulents or grass achieve a water surplus and they do not need to drink. On the other hand, desert herbivores like the oryx and Grant's gazelle are so adapted to the small amounts of water available from hygroscopic plants and rock crevices, that they also can live much of the year without drinking (Taylor, 1968b). Both sheep and kangaroos in moderately dry, cool environments can choose foods with sufficient water to enable them to live without drinking. In very dry, hot weather, however, they must find free water for evaporative cooling.

Most herbivores (other than seed-eaters) are too large to escape the rigours of the environment by taking advantage of ground shelter like burrowing rodents. Special adaptive mechanisms have evolved, therefore, for combating heat, cold, salt and aridity.

Under natural conditions most herbivores move considerable distances and expend energy while foraging. Scarcity of forage will force animals to walk greater distances in search of food. In desert, the available forage is meagre but the distribution of watering points is important to non-desert types. Water supplies interact with the physiological characteristics of the grazing animal and determine the effectiveness and evenness of utilization of arid pasture (Lange, 1969). This grazing zone centred on water is piosphere.

Claude Bernard emphasized that the internal environment is usually homeostatically regulated within narrow limits. So maintenance of water and salt balances by an animal depend in the long run upon equality of gain and loss.

The maintenance of an adequate osmotic extracellular environment

TABLE I

Water content and water turnover in various herbivores

Animal	Body wt (kg)	Total water % of body wt	Turnover of water ml kg^{-1} 24 h^{-1}	Conditions	Reference
Cattle					
Shorthorn	332	64.9	100.6	25°	Macfarlane and Howard (1972)
Shorthorn	332	73.9	163	36°	Macfarlane and Howard (1972)
Boran	197	77.0	136	37°	Macfarlane and Howard (1972)
Zebu	532	61.4	123	36°	Macfarlane and Howard (1972)
Buffalo	353	78	212.3	36°	Macfarlane and Howard (1972)
Sheep					
Merinos	41	55.9	68	25°	Macfarlane and Howard (1972)
German merinos	47.8	75	140	15°	Degen (1976)
German merinos (lactating)	42.7	77	177.3	15°	Degen (1976)
Awassi	35.4	73	123.6	15°	Degen (1976)
Awassi (lactating)	44.8	72	152.7	15°	Degen (1976)
Goats	40	69	95	37°	Macfarlane and Howard (1972)
Beduin goats	23	78	92.8	30°	Shkolnik et al. (1972)
Beduin goats (lactating)	20	84	187.4	30°	
Camel	656	68	30	25°	Macfarlane and Howard (1972)
Camel	520	72	61	37°	Macfarlane and Howard (1972)
Camel (lactating)	483	72.5	88.5	Alice Springs, Australia	
Red kangaroo	25.8	68.7	39.5	18.5°–33.3°	Denny et al. (1975)
Euro	24	73.1	39.4	18.5°–33.3°	Denny et al. (1975)

of cells is one part of homeostasis, and the sustenance of intracellular osmotic pressures is a cognate function. These two interrelated osmotic systems have fairly well fixed ranges; but the regulatory functions are more strained in some environments and among some mammals, than others.

The interaction of water, electrolytes, proteins and environment meet in osmotic adjustment at the surfaces, cell walls and regulator systems of animals, which includes hormones.

II. OSMOREGULATION OF MAMMALIAN HERBIVORES

In the evolutionary process a major step occurred when animals left water for the land, some 400 million years ago. A relatively impermeable integument was evolved for life in air rather than water. With that came regulation of the extracellular medium in which cells lived, while sustaining the traditional high K, low Na cellular composition. Later, in birds then among mammals, temperature regulation was added. Milk as a cell product, approximating the cells in composition and osmolality, was secreted to feed the young in dry habitats. The complexity of the brain increased to provide behavioural adjustments for living further away from water sources. Desert survival required the combination of behavioural tactics with cellular adaptation at biochemical and physiological levels. But mammals have come to occupy arid habitats remarkably well and the most successful of them have been herbivores and seed-eating rodents.

Osmotic pressure derives from dissolved molecules attracting water across semipermeable membranes. If the mammal is regarded as a complex set of surfaces, there is a wide range of osmotic pressures between the interstitial fluid which surrounds cells, and the external medium which forms the habitat of animals. The skin with its associated sweat, sebaceous, milk and lacrimal glands has contact with air which has a vapour pressure of 10–20 torr. The 1 or 2 mm thickness of keratinized cells forming the epidermal integument holds back interstitial fluid, normally at 300 mosmol concentration (5100 torr), so there is an outward diffusion gradient. The much thinner and more permeable membrane of the lungs is in contact with about 37 torr of water vapour pressure in the alveoli and surprisingly it constrains the interstitial fluid within a 1 to 2 μm thick sheet of membrane. But vapour diffuses into this space down a steep gradient. The other major surfaces of the body are the alimentary and renal tracts, which are fairly permeable but have much thicker cell layers (10–20 μm) than in the lungs. Gut fluid is really part of the external medium, but secretions are poured into it. The range of alimentary fluid osmotic pressures is from 250 to 350 mosmol l^{-1} in abomasum or ileum and less with caecum or colon (200–250 mosmol l^{-1}).

The membranes and cells of the renal tubules are exposed to concentrations varying with ecophysiologically diverse types of mammals. In animals derived from forest environments where water is plentiful, the range of osmolalities found along the nephron is from 300 mosmol l^{-1} at the glomerulus up to about 1200 mosmol l^{-1} in the collecting duct of pigs and humans. The loop of Henle and collecting duct fluids may reach 2600 mosmol l^{-1} in cattle; in sheep 3800, in dikdik 4500 and in desert *Notomys* a maximum of 9300 mosmol l^{-1} is reached, relative to a plasma concentration of 300 mosmol l^{-1} (Macfarlane and Howard, 1972; MacMillen and Lee, 1972; Maloiy, 1972; Sperber, 1944). The bladder also maintains a concentrated urine, relative to the interstitial fluid, though there is some exchange across the bladder wall if fluid is retained long in that organ. Exchange may be either inward or outward from the bladder.

Sweat glands secrete a fluid which is hypotonic to interstitial fluid, rarely above 100 mosmol l^{-1}. Sebaceous and mammary epithelia are thicker than those of sweat gland cells and they separate fluids with nearly the same tonicity as that of the interstitial milieu.

There are three main mechanisms by which the internal and external fluids are adjusted for osmoregulation. They are diffusion, active transport and hormone-regulated secretion.

A. DIFFUSION

Water molecules pass relatively slowly across the skin. In spite of the 5100 torr of osmotic pressure in the interstitial fluid which bathes the epithelium, with its vapour pressure of 757 torr, some water molecules break loose and pass out between the keratinized cells of the epidermis. The rate of this diffusional water loss (transpiration) is proportional to the rates of energy and water metabolism. In mammals with high rates of metabolic turnover, there is a high rate of epidermal permeability (Haines *et al.*, 1974; Macfarlane, 1976). This derives from the high rate of water turnover in mammals with rapid rates of metabolism. Presumably the turnover of water in the cells of the deeper layers of the epidermis is great enough to require a considerable amount of interstitial fluid between them, and this in turn allows the escape of water molecules through the hardened epidermis. Diffusion in the opposite direction also occurs. Immersion in fresh water results both in swelling of the skin and the uptake of fresh water into the interstitial fluid along a gradient which is usually at least 300:1. But the amounts are negligible.

The lung alveoli and the bronchi are much more readily permeable to water than the skin. Water molecules diffuse into the incoming ventilatory air to saturate at 35–37 torr vapour pressure (Fig. 1). The 2 μm membrane restrains the outflow of interstitial fluid into the alveolus—though pulmonary oedema does occur after chemical or high altitude damage to the membrane. Fluid with electrolytes at 300 mosmol l^{-1} concentration then passes into the alveolar spaces. Water and

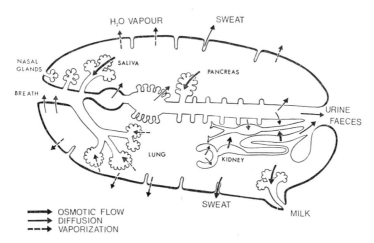

Fig. 1. Movement of water across surfaces in a mammal. Water moves passively by vaporization across skin and lung membranes. It diffuses in body fluids, and moves along osmotic gradients in gut, glands and kidney. These movements are part of the osmoregulatory processes. Secretion and hormones intervene to alter the electrolyte or crystalloid concentrations achieved by active transport. These determine the osmotic movement of water. Osmotic flow arises when assisted or active transport of solutes by secretory processes cause the solvent, water, to flow to osmotic equilibrium.

urea diffuse everywhere in a mammal. In small monogastrics, this is complete in 1 h, but large, slow-turnover mammals like the camel and the oryx may not equilibrate gut fluids with water and urea for 10 h.

Diffusion is also involved in the equilibration of some materials in the terminal part of the nephron. Urea diffuses back with water into the interstitial fluid and water itself diffuses across the collecting duct membrane towards the more concentrated interstitial fluid surrounding the loop of Henle. Resistance to this diffusion is reduced by the action of vasopressin working through cyclic AMP to change the apparent pore permeabilities of the duct plasma membrane (Orloff and Handler, 1967). These operations take place between the external world, the collecting duct lumen and the internal medium of interstitial fluid.

B. SECRETION

The secretion of sweat and tears provides a transient external milieu with lower osmotic pressure than that of the interstitial fluid. Sodium is reabsorbed from the fluid secreted into the duct, and at the end of the duct there is an osmotic gradient of 3 or 4:1, largely comprising K in ruminants, tending to draw water back into the body. There is a smaller gradient in the case of tears.

The muzzle glands of pigs and bovids are not sweat glands but they are multilobular acinar salivary glands (Brewer and Macfarlane 1967). They secrete a fluid which is modified by the duct, so that it contains considerably more K than Na. The glands are cholinergic in innervation and they secrete both mucus and electrolytes. In cattle the terminal fluid has an osmolality of about 150 mosmol l^{-1}. In this respect it is hypotonic to saliva and the secretions of stomach, pancreas, bile and intestine. All of these are poured on to the external surface—the alimentary tract—and there are ionic gradients (but only a small osmotic gradient) between them and the intercellular fluid.

The kidneys are major glands in the regulation of water for osmotic purposes. The regulation of extracellular fluid is based on Na control through modulation of the rate of Na excretion by aldosterone and deoxycorticosterone. Sodium concentration is monitored by the kidney cells, particularly the juxtaglomerular cells, while extracellular volume is registered by the vagal afferents of the atrium, acting through the brain stem on the renal nerves and thus on the release of renin. Sodium concentration and volume deficits lead to release of renin from juxtaglomerular cells and its enzymatic action on α globulin to produce angiotensin I. This is trimmed of two amino acids by a lung enzyme to produce the octapeptide angiotensin II. Renin and angiotensin II are present in parallel in ruminant plasma (Blair-West *et al.*, 1970). Angiotensin II in turn releases aldosterone from the glomerular layer of the adrenal cortex.

There are four main stimuli to the release of aldosterone (Blair-West *et al.*, 1963).

1. Low plasma Na or raised plasma K act directly on the adrenal.
2. Reduced volume of plasma, acting through the heart receptors to cause release of renin through the renal nerves.
3. Juxtaglomerular and macula densa responses to urinary Na concentration, and filtration pressures to activate the renin–angiotensin–aldosterone system.
4. Central processes involving ACTH and probably some mechanism independent of this hormone, activating the glomerular layer of the adrenal.

Among herbivores high K and reduction of plasma volume by dehydration or lack of Na are the most probable activations of aldosterone secretion.

Aldosterone encourages additional Na reabsorption by increasing enzyme synthesis in kidney, salivary glands, sweat glands and gut cells. In animals other than ruminants aldosterone increases K excretion, but sheep, cattle and camels (Macfarlane et al., 1967) do not respond by K secretion. Instead, these animals increase K secretion from the kidney under the influence of vasopressin (Kinne et al., 1961). Herbivores (which ingest the large amounts of K in plants) produce high concentrations of K from sweat glands and muzzle glands; and they eliminate K when water supplies are low, by renal action. This helps protect the circulation from the effects of hyperkalaemia.

Grazing on mountains or in the tropics where Na has been leached from the soil by heavy rainfall, yields low Na intakes (Howard et al., 1962). So mountain and tropical herbivores have high levels of aldosterone to retain Na. This works effectively so that only a few milliequivalents daily are needed to replace Na losses. There is no evidence that sustained high levels of aldosterone have deleterious effects.

The other side of renal osmoregulation in herbivores, is through control of water excretion. There are three mechanisms involved— filtration, proximal tubule Na–water reabsorption and collecting duct water uptake to interstitial fluid round the loop of Henle, under the influence of vasopressin.

The proximal tubule provides the essential regulation, with secondary control in the loop and collecting duct. There can be filtration control also. When a camel is without water for 10 days in summer (Siebert and Macfarlane, 1971; Maloiy, 1972) the filtration rate (GFR) may fall to one-fifth of normal, thus reducing urine flow. But Na and urea are retained (and tolerated) in the body for use later when water is ingested. A plasma osmolality of 450 mosmol l^{-1} is not disabling to camels.

The effectiveness of vasopressin is limited. At high flow rates vasopressin does not prevent urine flow (Macfarlane and Good, 1976) nor is it effective at very low rates of urine formation. Probably the lowest rates of urine flow result from glomerular shut-down and low GFR, on a vasopressin background. Sensitivity to vasopressin (within the usual range of urine flows) is low in ruminants like cattle, intermediate in sheep and goats and greatest in desert types like the camel, which is at least 50 times more reactive to vasopressin than cattle (Macfarlane et al., 1971).

When sheep are exposed to the sun and dry heat, panting reduces plasma water. Vasopressin is not readily released, however, and 2 to 3 days without water may pass before plasma vasopressin is significantly raised. This probably results from the mobilization of rumen water and the excretion of Na, which keeps down the plasma Na concentration. After 5 days without water in summer, Merino sheep averaged 157 mEq l^{-1} only for plasma Na (Macfarlane et al., 1956). Panting ruminants appear to excrete Na^+ during evaporative cooling, while sweating European men reabsorb renal Na almost completely.

Sheep produce vasopressin more readily in summer than in winter (Macfarlane et al., 1953) when exposed to the same thermal load. The induction of panting and sweating in sheep during winter did not lead to a rise of plasma vasopressin until deoxycorticosterone (DOC) was injected, then vasopressin rose to 180 μU ml^{-1} of plasma, about 100 times above the baseline concentration for sheep (Macfarlane and Robinson, 1957). The salt-active hormone (DOC) led to a reduction of interstitial volume during heating, and this appeared to cause vasopressin release. So vasopressin, aldosterone and seasonal changes in osmoregulatory function are linked together.

The regulation of water and electrolytes in grazing mammals is thus different, in relation to both aldosterone and vasopressin, from the glandular mechanisms of the kidney in non-ruminants like carnivores and omnivores (man, pig, dog, rat).

C. CELL OSMOREGULATION

The most obscure part of the osmoregulatory processes takes place at the cell membrane (Tosteson, 1964). Although cells have osmotic pressures of around 5100 torr, deriving from the electrolyte content (mainly K, Mg, Na, HCO_3, Cl and organic anions) there is also a protein onchotic pressure which may reach 70 torr. Membranes of all cells carry Na-K-activated ATPase (Skou, 1965) associated with Na transport outwards, activated from the inside, together with K intrusion from outside the cell. Most models of transport involve phosphorylation of the large protein component of the enzyme but not of the smaller glycoprotein moiety (Dahl and Hokin, 1974). Energy from ATP is made available for pumping Na out of cells and taking K up a 50:1 gradient into cells to maintain osmotic pressure and cell turgor. When the Na-K-activated ATPase is blocked by ouabain (Schwartz et al., 1975) there is no complete cessation of Na transport, and there is no swelling of cells in dilute media. Electrolytes pass freely across the membrane, and cells behave more like onchometers (in which differences of colloid osmotic pressure determine the water content of cells) rather than like osmometers working through the balance of electrolytes. The Na-K-activated ATPase is important for restoration of excitability in muscle, nerve or gland electrolytes after action, and for the work of gut in absorption or kidney in secretion and reabsorption; but it is evidently secondary to the cell proteins in determining the osmotic equilibrium of cells.

The activity of Na–K-activated ATPase is increased by thyroxine and some of the energy release due to thyroxine appears (Ismail-Beigi and Edelmann, 1971) to be the result of greater Na pumping. There is not usually any osmotic consequences of greater pump action. Cortisol in ruminants increases the Na–K-activated ATPase activity of cells in the kidney, and is more involved with this than aldosterone (which does not affect the enzyme).

As with water, electrolyte balances in animals are subject to central regulatory processes and are kept relatively constant on a day-to-day basis. There are some storage and tolerance components in the system also. Some Na is stored in bone, while there is a reserve of water and K in the reticulo-rumen of ruminants. And the Na space is expanded during the hot season (Macfarlane *et al.*, 1959). The components of osmoregulation may now be looked at in relation to specific adaptations. Water handling is one aspect of this, and electrolytes are the other major part of the process.

1. *Water Balance*

Any living organism is analogous to an open flow-through system whereby there is a continuous exchange of material and energy with the environment. In this context, the exchange and turnover of water exceeds that of any other constituents. The avenues by which an animal gains and loses water are summarized in Fig. 2.

There are daily imbalances in the system, and these may persist or fluctuate over a range of 20 to 30% through the seasons.

The fluids within the animal body are not continuously distributed. Several compartments differing both in volume and ionic composition are distinguishable (Fig. 3). In different herbivores the relative volume of the compartments differ greatly and are differently affected by heat and dehydration.

The avenues by which a herbivore takes in water are depicted in Fig. 2. The amount of water relative to the total intake a herbivore may drink depends on the water content of its food and very often on the availability of surface water present in its habitat.

Preformed water in the food varies greatly with the nature of the animal's diet. In swamps the plant material may be 85% water. Some dry desert fodder eaten by the oryx and Grant's gazelle has been shown (Taylor, 1968b) to be hygroscopic and its water content changes with the relative humidity as well as the temperature of ambient air. These

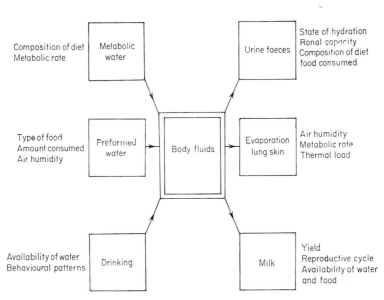

Fig. 2. Water balance in a herbivore. Avenues of water intake and water loss.

variables fluctuate greatly between day and night in the desert environment. Dew often forms at night in the desert and plant material can hold this water till the sun dries it.

Metabolic water on the other hand is affected only by the nature and composition of the food. Fats and oils, mainly in seeds, yield twice as much water from oxidized hydrogen as do sugars or proteins from leaves and stems.

Free drinking water is imbibed to differing degrees by herbivores of different ecophysiological origin. Some wet country animals like the Cape buffalo (*Syncerus caffer*) and others cattle; waterbuck (*Kobus ellipsiprymnus*); Sitatunga (*Limnotragus spekei*) and the horse, to mention only a few, prefer to drink water daily and are always found in the vicinity of water sources or within a day's walk from a watering point. Others like the oryx, camel, gazelle or dikdik which evolved for the desert, can live with little or no drinking water despite the fact that they inhabit hot, arid environments. Sheep and kangaroos in temperate climates, with 55 to 75% of water in pasture plants, do not need to drink. Oryx and addax antelope in the desert also seem to go through considerable part of the year without drinking. Intermediate types (sheep, goats, wildebeest or kongoni) drink each 2 to 3 days during hot, dry season.

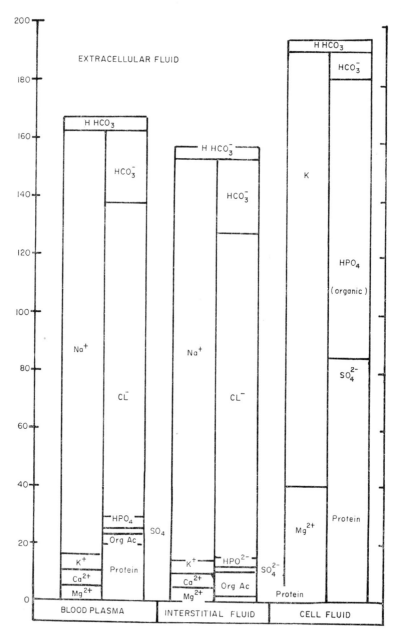

Fig. 3. The ionic composition of plasma, interstitial fluid and intercellular fluid (mEq l^{-1}). The figures to the left of the ordinate refer to concentration in mEq l^{-1} H$_2$O. The figures on the right indicate the sum of the concentration of anions and cations. The main cation in extracellular fluid is Na$^+$ and the chief anion is Cl$^-$. In intracellular fluid K$^+$ predominates and there are large amounts of protein and organic phosphates ions. Note the low concentration of protein in interstitial fluid (after Gamble, 1964).

Figure 4 depicts the mean body composition and body fluid distribution in both normally hydrated and dehydrated herbivores.

The minimum amounts of water required by different herbivores subjected to heat in a climatic chamber are shown in Table II.

Herbivores lose water by urine, faeces, evaporation and through milk. Table III gives the amount of water expended for evaporative cooling in both dehydrated and hydrated antelopes.

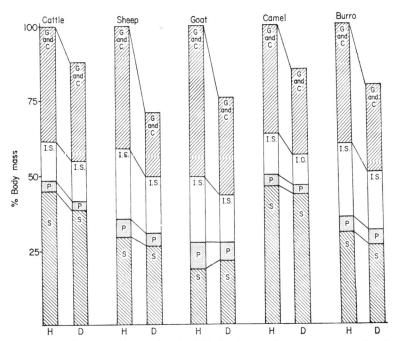

Fig. 4. Mean body composition and body fluid distribution in normally watered (H) and water-deprived (D) herbivores (S: body solids; P: blood plasma; ISF Interstitial fluids; G & C: gut contents).

In hot climates the major avenue of water loss is through evaporation. This is effected through the respiratory passages (lungs, turbinates), through the skin by diffusion, or in animals over 50 kg, by sweating.

Gaseous exchange is continuous in the lung alveoli. Water in turn is evaporated from the respiratory passages during ventilation.

Evaporation from the respiratory passages is dependent on the amount of air per unit time passing over the moist respiratory passages. Increase in convection is usually achieved by an increase in respiratory frequency and a decrease in depth of breathing. During vigorous exercise and following exposure to a high thermal load, most herbivores dissipate excess heat produced in the body by panting and sweating. Smaller

TABLE II

Minimum water requirements in herbivores exposed to heat

Species	l H_2O 100 kg body wt^{-1} 24 h^{-1}
Hereford	6·42
Zebu	3·22
Buffalo	4·58
Eland	5·49
Wildebeest	4·81
Oryx	3·00
Thomson's gazelle	2·74
Grant's gazelle	3·86
Impala	2·98
Hartebeest	4·02

The animals were fed on hay (*Cynodon*) and exposed to 40°C for 12 h each day for several days. Minimum water requirements were assessed in terms of a standardized body weight. Data from Taylor (1968a); Maloiy and Hopcraft (1971).

TABLE III

Water loss from herbivores by evaporation

Species	Normal hydration		Dehydration	
	22°C	22–40°C	22°C	22–40°C
Grant's gazelle (*Gazella granti*)	2·74 ± 0·18	4·68 ± 0·22	1·41 ± 0·07	3·25 ± 0·06
Thomson's gazelle (*Gazella thomsoni*)	3·65 ± 0·31	6·50 ± 0·28	1·71 ± 0·10	2·38 ± 0·04
Oryx (*Oryx beisa*)	2·01 ± 0·22	3·74 ± 0·23	0·91 ± 0·07	2·15 ± 0·12
Wildebeest (*Connochaetes taurinus*)	1·36 ± 0·14	4·59 ± 0·30	1·10 ± 0·09	3·34 ± 0·07
Eland (*Taurotragus oryx*)	1·89 ± 0·29	4·53 ± 0·24	1·63 ± 0·12	3·56 ± 0·24
Buffalo (*Syncerus caffer*)	1·81 ± 0·14	4·10 ± 0·19	1·23 ± 0·09	3·07 ± 0·09
Zebu steer (*Bos indicus*)	1·33 ± 0·19	4·24 ± 0·25	0·85 ± 0·05	2·10 ± 0·09
Hereford steer (*Bos taurus*)	2·97 ± 0·27	4·70 ± 0·31	2·14 ± 0·16	3·83 ± 0·29
Hartebeest (*Alcelaphus buscelaphus*)	2·60 ± 0·23	5·97 ± 0·18	1·45 ± 0·08	3·16 ± 0·13
Impala (*Aepyceros melampus*)	2·25 ± 0·14	3·94 ± 0·24	1·34 ± 0·06	1·92 ± 0·07
Fat-tailed sheep (*Ovis aries*)	3·99 ± 0·51	6·59 ± 0·62	1·96 ± 0·12	3·12 ± 0·04
Turkana goat (*Capara hircus*)	3·79 ± 0·79	6·45 ± 0·32	2·35 ± 0·07	3·24 ± 0·11

Figures are given in l 100 kg body wt day^{-1} ± s.e.
Data are from Taylor (1970); Maloiy and Hopcraft (1971); Maloiy and Taylor (1971).

herbivores of up to 40 kg body mass usually pant without sweating on exposure to heat, while a majority of larger ones dissipate excess heat by sweating (Fig. 5). Many large herbivores have both mechanisms (sheep, cattle, eland).

In the panting category are included such herbivores as the hyrax, dikdik, suni antelope and gazelles. Characteristic examples of the sweating group of herbivores are horses, donkeys, eland, camels and buffalo. Sweat glands play an important role when augmented evaporative cooling becomes essential. The secretory functions of apocrine sweat glands in herbivores are under the control of cholinergic neurochemical transmitters (Table IV) either from sympathetic nerves or by adrenaline directly inducing secretion.

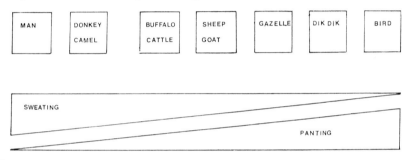

Fig. 5. Relative importance of sweating and panting in various species. The diagram depicts the general trend and not the exact quantitative relationship of thermal sweating and panting.

TABLE IV

Stimulation and control of sweat gland activity in herbivores

Species	Response to heat	Pharmacological response	
		Adrenaline	Acetylcholine
Equine	+	+	+
Camel	+	+	−
Cattle	+	+	−
Sheep	+	+	−
Goat	+	+	−
Wildebeest	+	+	−
Eland	+	+	−
Buffalo	+	+	−
Oryx	+	+	−
Dikdik	−	+	−
Grant's gazelle	+	+	−
Thomson's gazelle	+	+	−

Data from Jenkinson (1972); Maloiy (1973).

Sweat glands have been found histologically to be present in the skin of such herbivores as the buffalo, camel, cattle, horse, donkey, goat, sheep and several antelope species. The density of these sweat glands however, varies greatly from one species to another. For example, in dikdik, goats, sheep and camels there are 200–400 sweat glands cm^{-2} while in some zebu cattle there are 2000 glands cm^{-2} (Jenkinson, 1972).

The renal system constitutes another avenue of water loss, and the kidney is important in controlling salt levels for osmoregulation. Under conditions of normal hydration, excess water and metabolic waste products are excreted by the kidney. Dehydration on the other hand

TABLE V

Comparison of osmolality and U/P ratio in dehydrated ruminants and the Somali donkey

Animal	Osmolality (mosmol kg H_2O^{-1})	U/P osmolal ratio	Reference
Dikdik	4100	11	Maloiy (1972)
Awassi	3230	8·3	Degen (1976)
Camel	3200	8·0	Maloiy (1972)
Merinos	3200	7·3	Macfarlane et al. (1956)
Oryx	3100	8	Taylor (quoted by Maloiy, 1972)
Fat-tailed sheep	3100	8	Maloiy (1972)
Goat	2900	8	Maloiy (1972)
German merino	2896	7·6	Degen (1976)
Impala	2250	6·3	Maloiy and Hopcraft (1971)
Beduin goat	2200	7·0	Choshniak (1972)
Hartebeest	2010	5·7	Maloiy and Hopcraft (1971)
Eland	1881	6	Taylor and Lyman (1967)
Somali donkey	1680	4·7	Maloiy (1970)
Zebu	1300	4	Taylor (quoted by Maloiy, 1972)
Hereford	1160	4	Taylor and Lyman (1967)
Wildebeest	1832	6	Maloiy and Taylor (unpublished)
Grant's gazelle	2789	8	Maloiy and Taylor (unpublished)
Rock hyrax	3180	8	Maloiy and Taylor (unpublished)
Buffalo	1124	4	Maloiy and Taylor (unpublished)
Thomson's gazelle	2638	7	Maloiy and Taylor (unpublished)

is accompanied by a decrease in urine volume and a concomitant rise in extracellular osmotic pressure. The kidneys are brought into play first to save fluid and second to adjust the electrolyte status.

Under conditions of normal hydration, most herbivores excrete a urine which may vary greatly in volume and its ionic and osmotic concentration. In the hyrax and dikdik for example, urine volume may range from 100 to 300 ml day^{-1} when the animals are fully hydrated. The volume is, however, reduced to 15–60 ml day^{-1} following one or two days of water deprivation. Similarly in Beduin and African goats, urine volume amounts to 400–550 ml day^{-1} when these animals

TABLE VI

Faecal water content of dehydrated herbivores

Animal	% H$_2$O in faeces	Reference
Sinai goat	40·0	Shkolnik et al. (1972)
Camel	43·2–44·1	Maloiy (1972), Schmidt-Nielsen (1964)
Dikdik	44·0	Maloiy (1972)
German Merino	44·3	Degen (1976)
Awassi	45·0	Degen (1976)
Somali fat-tailed sheep	48·2	Maloiy and Taylor (1971)
Turkana goats	51·5	Maloiy and Taylor (1971)
Hartebeest	51·9	Maloiy and Hopcraft (1971)
Impala	53·3	Maloiy and Hopcraft (1971)
Somali donkey	60·9	Maloiy (1970)
Eland	61·5	Taylor and Lyman (1967)
Hereford	75·1	Taylor and Lyman (1967)

are watered regularly. Dehydration reduces this volume by half. Similar reduction in urinary water loss values for the camel, donkey, sheep, cattle, eland and other dehydrated herbivores have also been observed.

These observed changes in urine volumes in herbivores are the net result of either changes in GFR or renal tubular water reabsorption or both. Renal concentrating ability and osmolal U/P ratio are depicted in Table V for a variety of herbivores.

Water is reabsorbed throughout all regions of the gastrointestinal tract. The final conservation of water which takes place before faeces are egested occurs in the large intestine, especially in the caecum and the colon. In these regions of the gut, water reabsorption is achieved as Na is actively absorbed. The amount of water lost through this route (Table VI) is a function of the amount of dry faecal matter, the residual water (Table VI) as well as of the habitat. Buffalos in swamps have wet faeces (80% water), cattle also excrete wet faeces, but goats, giraffe, camels and dikdiks form pellets with only 40–50% of water.

The water metabolism of herbivores is affected by the costs of milk secretion. The effect of milk production on the water turnover of a few herbivores is given in Table I. The elevated water turnover in lactating animals (40 to 50% above normal) reflects not only the water content of the milk but higher water demands of the increased metabolism of the milk-secreting animal (Macfarlane and Howard, 1972; Meltz, 1975).

2. *Electrolyte Balance*

Substantial quantities of electrolytes are taken in by herbivores through the food. In the absence of fresh water, however, some herbivores may utilize salt waters or eat salt-storing chenopods like *Atriplex* and *Kochia*. Several studies have been carried out on the salt tolerance of herbivores. The donkey (Maloiy, 1972) sheep and cattle (Macfarlane, 1971) tolerate concentrations of 180–220 mmol l^{-1} NaCl in their drinking water. Turkana goats are not affected by higher concentrations (250 mmol l^{-1}) and the concentration tolerated by Black Beduin goats was 500 mmol l^{-1}. Drinking of sea water (1000 mmol l^{-1}) on the other hand, has been claimed by Dunson (1974) to improve the water economy of the goats that inhabit dry areas in the Galapogos Archipelago.

Maloiy (1972) found the camel to be unusual among herbivores in its salt tolerance. The highest concentration of salt in drinking water ingested by this animal with no ill effects was about 1850 mmol l^{-1} (5·5% NaCl). Even with this high salt intake, urine osmolality did not rise above 3000.

The ability of an animal to use salt solutions or sea water is an adaptive characteristic best developed in desert herbivores. Surface water is scarce in the desert and pools, becoming more salty by evaporation, may be the only drinking water available. Evolutionary pressures probably arose from this, so that cell pumps could cope with high Na concentrations.

Evaporation of water from the respiratory tract is usually accompanied by a corresponding loss of ions through the kidney to excrete residual Na (Macfarlane *et al.*, 1958). When cattle and similar animals are hot, drooling of saliva normally takes place in herbivores as a result of thermal panting. This process results in a loss of both fluid and electrolytes. Although much data is available on the relative concentration of ions and rate of secretion of saliva (Table VII) in such herbivores as the ox, sheep, goat, llama, camel and horse (Kay, 1960; Hoppe *et al.*, 1974), little is known about the saliva secretion rate during cooling.

Of interest here is the discovery by Steno, the Danish anatomist, in

TABLE VII

Composition of parotid saliva

Species	Osmolality (mosmol kg H_2O)	Na^+	K^+	Total CO_2 mEq l^{-1}	HPO_4^{2-}	Cl^-
Camel	283	152	24	119	8	16
Sheep	—	170	13	112	48	11
Calves	—	166	10	—	32	25

From Hoppe et al. (1974); Kay (1960).

1664 of a gland situated in the wall of the maxillary sinus, opening into the nasal sinus in sheep. Recent studies by Johnson and Peaker (1974) aimed at assessing the role of this gland in evaporative cooling in sheep have revealed interesting results. The rate of secretion of the gland increased from 0·003 to 0·207 ml h^{-1} gland^{-1} upon exposure at a T_a 45°C for 2–3 h. Secretion is hypotonic to blood 30–135 mosmol kg H_2O^{-1}. The concentration of Na under these conditions ranged from 8 to 79 mEq l^{-1} and that K was 5–16 mEq l^{-1} while Cl was 16–81 mEq l^{-1}.

The part taken by this gland in evaporative cooling in the dog has been demonstrated by Blatt et al. (1972). But there appears to be no information on the functions or the anatomical existence of this gland in other herbivores.

Rates of cutaneous evaporation have been measured in the range of 50 to 650 g m^{-2} h^{-1} among sweating herbivores. Potassium, and less Na with HCO_3 accompany this elimination of water through the skin glands of herbivores. Both ventilated and unventilated sweat capsules applied to a closely shaven area of the skin have been employed to measure the rate of cutaneous water loss in herbivores (McLean, 1963). The physical as well as the analytical procedures involved in either of these methods as well as their practical limitations in estimating sweating rate have been described. To date, however, no one standardized method or technique is available for either collecting or measuring the ionic composition of the fluid secreted during sweating in herbivores. Most authors (Joshi et al., 1969; Johnson, 1969; Jenkinson and Mabon, 1972) have measured the composition of sweat of cattle from samples collected on filter papers which have been enclosed in unventilated sweat capsules, which in turn have been attached tightly to closely shaven areas of the animal skin. Washed out samples from these areas are then utilized for the quantitative analysis of their constituents. This method has its practical limitations and is indeed unsatisfactory. It is the source of variable but contradictory results so far reported in the literature.

In Zebu cattle exposed to a T_a 45°C, Johnson (1969) found K concentrations in cutaneous secretions to approximate 400 μmol m^{-2} h^{-1} and those of Na to be about 100–150 μmol m^{-2} h^{-1}. These ionic losses amount to 1–3% of the total K and Na ingested in the feed. In the same study, Johnson found lower K$^+$ and Na$^+$ concentrations in the cutaneous secretion of British breeds of cattle that he investigated under identical conditions. No significant losses of these ions were observed in these breeds.

In other studies, again on cattle, higher losses of K and Na in cutaneous secretions have been measured, than those reported by Johnson (1969). A K/Na ratio of 4–5 was reported, however, in all studies (Freeborn et al., 1934; Taneja, 1956; Stacy et al., 1963). The differences in these results could partly be due to the differences in the experimental methods in sampling and partly by the lack of standardization of the experimental design—such as the length of exposure to high ambient temperatures.

In the sweat of Indian breeds of cattle exposed to high ambient temperatures, Joshi et al. (1968) found considerable concentrations of inorganic phosphorus, nitrogen, chloride, reducing sugars and lactic acid. Proteins were also found in cattle sweat by these authors and they are present in horse sweat.

In a study on sheep sweat, Stacy et al. (1963) found the K/Na ratio to range between 3 and 8. The cutaneous secretion of Na and K amounted to 2·8% and 1·4% respectively of the amount of those ions ingested by sheep. It appears then, that on the basis of these results that little Na is normally lost by herbivores through sweating. Observations by Schmidt-Nielsen (1964) on the donkey and horse and by Macfarlane (1964) on sheep and camel support this view. In the camel sweat for example, Macfarlane (1964) found that Na and K concentrations were between 9 and 40 mEq l^{-1} respectively.

3. Milk and Osmotic Balance

Milk secreted by acinar glands derived from the ectoderm, is essentially isoosmotic with plasma (Rook and Wheelock, 1967). Although the total solids content of milk range from 10% in man to 20% in sheep and over 50% in whales, the electrolyte component is less variable. In cows' milk there are three times more Na than K and low salt intake results in less Na and more K appearing in the milk. Since milk is holocrine cellular secretion, there is Mg, SO_4, PO_4 in solution, as well as Ca.

Lactose and fats contribute to the osmotic pressure, but dissociating electrolytes are the major osmotic components. The osmotic drain on the lactating female can be considerable, when both water and solutes equal to 3 to 15% of body weight are drained off each day. There is particularly large drain on Ca, which draws on bone reserves if food Ca is deficient. And in highly productive dairy animals, the Ca loss may lead to hypocalcaemia in the context of pregnancy toxaemia or milk fever. Replacement of calcium by intravenous infusion is often sufficient to restore disordered metabolism (Kronfeld, 1969).

4. *Mineral Deficiency*

Herbivores automatically ingest sufficient K for cellular and extracellular functions, when they are feeding on plants which have ample K for animal needs.

There is no equivalent of the animal's extracellular Na space in plants, so that in leached areas of heavy rainfall, or in the Arctic, there is often little Na in plants. Tropical mountain pastures comprise plants with the lowest Na content, where the Na may be only one-tenth that of temperate zone plants (Howard *et al.*, 1962). Hutton (1958) has shown that Na in soil and plants decreases with distance from the sea (unless there is an ancient sea bed plain as substrate).

Both men (Macfarlane, 1968, 1976) and other animals adjust well to low Na intake since aldosterone circulates in large quantity and reduces Na in sweat, saliva, urine and milk. In humans, blood pressures remain low among vegans on 5–30 mEq day^{-1} Na intake. No disability is known to arise in New Guinea vegans from their low Na intake, nor is there significant evidence that ruminants are adversely affected by the adjustments to low Na. Jordan *et al.* (1972) and Scroggins *et al.* (1970) have reported on the aldosterone response of herbivores for Na conservation. Osmotic balance is maintained with extracellular Na at 135–145 mEq l^{-1} in spite of small intakes of Na and high plasma aldosterone concentrations.

5. *Functional Hierarchies*

Evidence is accumulating that present-day genera of mammals achieved their current levels of function 5–10 million years ago. Evolution in a given habitat resulted in the fixing of a set of rate functions, which were optimal for the food, water and climate conditions of the Pliocene. There probably has been little functional change since (Simpson, 1953). Yet there were great migrations of mammals across latitudes, as well as between the Americas and Asia in both directions during the Pleistocene. Many animals survived in ecosystems very different from those of their basic evolutionary earth. Thus wet tropical cattle have been moved to arid grasslands; deer from wet temperate regions have come to live in the dry Arctic with high rates of water and energy use like the reindeer and moose; while camels and llamas, parted 3 million years ago in the south-west Rockies region, have retained low energy and water-use patterns in the very different environments of the Sahara, the Gobi and the Andes or Patagonia (Macfarlane, 1976).

Size is linked exponentially to energy use rates ($kg^{0.75}$) though there is a range of 300% between the highest and lowest oxygen use rates at a given size. Similarly, for water use the average relation is $kg^{0.82}$, but cattle and camels of the same mass differ by over 200% in water turnover, even when they eat the same food in the same environment (Macfarlane and Howard, 1972).

Mammalian herbivores may be divided into three main physiological ecotypes in terms of their water and salt-handling functions. With those essentials of osmoregulation go other functions such as renal concentrating power, response to vasopressin, and the rate of protein turnover. Tolerance to excess Na, and Na–K-activated ATPase activities may also be linked to the water, energy, protein turnover rates. The main rate function groups are listed below.

1. Wet tropical or wet temperate origins, with high rates of water and energy use, poor renal concentration and low salt tolerance—buffalo, cattle, pig, eland, suni, waterbuck, elephant, horse, moose, reindeer.
2. Warm dry savannah, semi-arid. Intermediate rates of water and energy use, good urine concentration, moderate salt tolerance—sheep, wildebeest, kongoni, hyrax, donkey.
3. Arid zone animals with low rates of energy and water turnover, but tolerant of salt, and with medium to high urine concentration—camel, goat, oryx, *Dipodomys*, *Notomys*, gazelles.

The camel reduces filtration rather than concentrate urine above 3100 mosmol l^{-1} but *Dipodomys* and *Notomys* achieve two or three times that concentration.

In all groups cellular and extracellular osmolalities remain much the same. It is when exposed to lack of water or excess of salt that the essential differences in adaptation become apparent. The wet tropical group die within 4–5 days without water in summer and lose co-ordination when drinking 1% NaCl.

In contrast camels in the desert group may live 15 or more days without water dry season and are not poisoned by 5% NaCl in drinking water. These are part only of a complex of rates of using water, energy and protein, as well as of Na pumping. The renal tubules have longer loops of Henle in desert-adapted forms and vasopressin is more active in the reabsorption of water. Faeces are dried, also, to a greater extent than among the higher turnover rate animals. The total complex of linked adaptive functions includes behaviour, and all the features appear to have evolved pretty much as a package. There are some deviations from standard in each group, as among the camels with lower urine concentration maximum than sheep, but with better summer survival without water.

REFERENCES

Blair-West, J. R., Coghlan, J. P., Denton, D. A. and Wright, R. D. (1967). Effect of endocrines on salivary glands. In "Handbook of Physiology. Alimentary Canal." Washington D.C., Am. Physiol. Soc. Vol. 2, pp. 633–664.

Blair-West, J. R., Coghlan, J. P., Denton, D. A. and Wright, R. D. (1970). Factors in sodium and potassium metabolism. In "Physiology of Digestion and Metabolism in the Ruminant" (A. T. Phillipson, ed.). Oriel Press, Newcastle, pp. 350–361.

Blatt, C. M., Taylor, C. R. and Habel, M. B. (1972). Thermal panting in dogs: The lateral nasal gland, a source of water for evaporative cooling. Science, N.Y. 177, 804–805.

Chosniak, I. (1974). The water economy of the black Bedouin goat. M.Sc. Thesis, Tel-Aviv University, Israel.

Brewer, N. L. and Macfarlane, W. V. (1967). Structure and secretion of the nasolabial glands of cattle. Aust. J. exp. Biol. med. Sci. 45, 37.

Dahl, J. L. and Hokin, L. E. (1974). The sodium–potassium adenosinetriphosphatase. Ann. Rev. Biochem. 43, 327–356.

Denton, D. A. (1965). Evolutionary aspects of the emergence of aldosterone secretion and salt appetite. Physiol. Rev. 45, 254–295.

Degen, A. A. (1975). Physiological Responses of Two Breeds of Sheep in a Desert. Ph.D. Thesis, Tel-Aviv University, Israel.

Denny, M. J. S. and Dawson, T. J. (1975). Effects of dehydration on body water distribution in desert kangaroos. Am. J. Physiol. 229, 251–254.

Dunson, W. A. (1974). Some aspects of salt and water balance of feral goats from arid islands. Am. J. Physiol. 226, 662–669.

Freeborn, S. R., Regan, W. M., Berry, L. S. (1934). The effect of petroleum oil fly sprays on dairy cattle. J. Econ. Entom. 27, 382–383.

Gamble J. L. (1964). "Chemical Anatomy, Physiology and Pathology of Extracellular Fluid." Harvard University Press, Cambridge, USA.

Haines, H., Macfarlane, W. V., Setchell, C. and Howard, B. (1974). Water turnover and pulmocutaneous evaporation of Australian desert dasyurids and murids. Am. J. Physiol. 227, 958–963.

Hoppe, P., Kay, R. N. B. and Maloiy, G. M. O. (1974). Salivary secretion in the camel. J. Physiol. (Lond.) 244, 32–33P.

Howard, D. H., Burdin, H. L. and Lampkin, G. H. (1962). Variation in the mineral and crude protein content of pastures at Muguga in Kenya Highlands. J. agric. Sci. Camb. 59, 251–256.

Ismail-Beigi, F. and Edelman, I. S. (1970). The mechanism of thermogenesis of thyroid hormones. *J. gen. Physiol.* **57**, 710–719.

Jenkinson, D. McE. (1972). Evaporative temperature regulation in domestic animals. *Symp. zool. Soc. Lond.* **31**, 345–356.

Jenkinson, D. McE. and Mabon, R. M. (1973). The effect of temperature and humidity and skin surface pH and the ionic composition of skin secretions in Ayrshire cattle. *Brit. Vet. J.* **129**, 282–295.

Jenkinson, D. McE. (1973). Comparative physiology of sweating. *Brit. J. Dermatol.* **88**, 397–406.

Johnson, K. G. (1970). Sweating rate and the electrolyte content of skin secretions of *Bos taurus* and *Bos indicus* cross-bred cows. *J. agric. Sci. Comb.* **75**, 397–402.

Johnson, K. G. and Peaker, M. (1974). Studies of the lateral nasal (Steno's) glands of the sheep. *J. Physiol. (Lond.).* **242**, 7p–8p.

Jordan, P. A., Botkin, D. B., Dominski, A. S., Lowendorf, H. S. and Bleovsky, G. E. (1972). "Sodium as a Critical Nutrient for the Moose of Isle Royal." 9th North American Moose Symp., Quebec, Canada.

Joshi, B. C., Joshi, H. B., McDowell, R. E. and Sandu, D. P. (1968). Composition of skin secretions from three breeds of cattle under thermal stress. *J. Dairy Sc.* **51**, 917–920.

Kay, R. N. (1960). The rate of flow and composition of various salivary secretions in sheep and calves. *J. Physiol. (Lond.)* **150**, 515–537.

Kinne, R., Macfarlane, W. V. and Budtz-Olsen, O. E. (1961). Hormones and electrolyte excretion in sheep. *Nature, Lond.* **193**, 1084–1085.

Kronfeld, D. S. (1970). Ketone body metabolism, its control and its implications in pregnancy toxaemia, acetonaemia and feeding standards. *In* "Physiology of Digestion and Metabolism in the Ruminant" (A. T. Phillipson, ed.). Oriel Press, Newcastle, pp. 566–583.

Lange, R. T. (1969). The piosphere: sheep track and dung patterns. *J. Range Management* **22**, 396–400.

Macfarlane, W. V. (1971). Salinity and the whole animal. *In* "Salinity and Water Use" (T. Talsma and J. Philip, eds). Macmillan, London, pp. 161–178.

Macfarlane, W. V. (1976). Ecophysiological hierarchies. *Israel J. med. Sci.* **12**, 723–731.

Macfarlane, W. V. and Good, B. F. (1976). Hormones and adaptation. *In* "Selected Topics in Environmental Biology" (B. Bhatia, C. S. Chhina and B. Singh, eds). Interprint Publications, New Delhi, pp. 211–216.

Macfarlane, W. V. and Howard, B. (1972). Comparative water and energy economy of wild and domestic mammals. *Symp. zool. Soc. Lond.* **31**, 261–294.

Macfarlane, W. V. and Robinson, K. W. (1957). Seasonal changes in plasma antidiuretic activity produced by a standard heat stimulus. *J. Physiol. (Lond.)* **135**, 1–11.

Macfarlane, W. V., Robinson, K. W., Howard, B. and Kinne, R. (1958). Heat, salt and hormones in panting and sweating animals. *Nature, Lond.* **132**, 572–573.

Macfarlane, W. V., Morris, R. J. H., Howard, B. and Budtz-Olsen, O. E. (1959). Extracellular fluid distribution in tropical Merino sheep. *Aust. J. agric. Res.* **10**, 269–286.

Macfarlane, W. V., Kinne, R., Walmsley, C., Siebert, B. D. and Peter, D. (1967). Vasopressins and the increase of water and electrolyte excretion by sheep, cattle and camels. *Nature, Lond.* **214**, 979.

Macfarlane, W. V., Howard, B., Scroggins, B. and Skinner, S. L. (1968). Water electrolytes, hormones and blood pressure of Melanesians in relation to European contact. *Proc. XXIV Int. Cong. Physiol.* **7**, 274.

Macfarlane, W. V., Howard, B., Haines, H., Kennedy, P. M. and Sharpe, C. M. (1971). Hierarchy of water and energy turnover in desert mammals. *Nature, Lond.* **234**, 483–484.

McLean, J. A. (1963). The regional distribution of cutaneous moisture vaporization in Ayrshire calf. *J. agric. Sci. Camb.* **61**, 275–280

MacMillen, R. E. and Lee, A. K. (1969). Water metabolism of Australian hopping mice. *Comp. Biochem. Physiol.* **28**, 493–514.

Maloiy, G. M. O. (1970). Water economy of the somatic donkey. *Am. J. Physiol.* **219**, 1522–1527.

Maloiy, G. M. O. (1972). Renal salt and water excretion in the camel (*Camelus dromedarius*). *Symp. zool. Soc. Lond.* **31**, 243–259.

Maloiy, G. M. O. (1973). The water metabolism of a small East African antelope: the dik dik. *Proc. R. Soc. Lond. Ser. B* **184**, 167–178.

Maloiy, G. M. O. and Hopcraft, D. (1971). Thermoregulation and water relations in two East African antelopes: the Hartebeest and Impala. *Comp. Biochem. Physiol.* **38A**, 403–412.

Maloiy, G. M. O. and Taylor, C. R. (1971). Water requirements of African goats and haired sheep. *J. agric. Sci. Camb.* **77**, 203–208.

Meltz, E. (1975). Milk Production and Water Economy in the Bedouin Goat. M.Sc. Thesis, Tel-Aviv University, Israel.

Orloff, J. and Handler, J. (1967). The role of adenosine 3–5-phosphate in the action of antidiuretic hormone. *Am. J. Med.* **42**, 757–768.

Robinson, J. R. (1964). Renal handling of salt and water. *Aust. Ann. Med.* **13**, 183–191.

Rook, J. A. F. and Wheelock, F. (1967). The secretion of water and of water soluble constituents in milk. *J. Dairy Res.* **34**, 273–287.

Schmidt-Nielsen, K. (1964). "Desert Animals: The Physiological Problems of Heat and Water." Oxford University Press, Oxford.

Schwartz, A., Lindenmeyer, G. E. and Allen, J. C. (1975). The sodium–potassium triphosphatase: pharmacological, physiological and biochemical aspects. *Pharmac. Rev.* **27**, 3–115.

Shkolnik, A., Borut, A, and Chosniak, I. (1972). Water economy of the Bedouin goat. *Symp. Zool. Soc. Lond.* **31**, 229–242.

Simpson, G. G. (1953). The Major Features of Evolution. Columbia University Press, New York.

Skou, J. C. (1965). Enzymatic basis for active transport of Na^+ and K^+ across cell membrane. *Physiol. Rev.* **45**, 596–617.

Sperber, I. (1944). Studies on the mammalian kidney. *Zool. Bidrog. Fran Uppsala* **22**, 249–432.

Taneja, G. C. (1959). Sweating in cattle. IV. Control of sweat glands secretion. *J. agric Sci.* **52**, 66–71.

Taylor, C. R. (1968a). The minimum water requirements of some East African bovids. *Symp. zool. Soc. Lond.* **21**, 195–206.

Taylor, C. R. (1968b). Hygroscopic food: a source of water for desert antelopes? *Nature, Lond.* **219**, 181–182.

Taylor, C. R. (1970). Strategies of temperature regulation, effect on evaporation in East African ungulates. *Am. J. Physiol.* **219**, 1131–1135.

Taylor, C. R. and Lyman, C. P. (1976). A comparative study of the environmental physiology of an East African antelope, the eland and the Hereford steer. *Physiol. Zool.* **40**, 280–295.

6. Primates

L. P. SULLIVAN

University of Kansas Medical Center, Kansas City, Kansas, USA

I.	Introduction	211
II.	The Role of the Kidney in Osmoregulation	212
	A. Structure of the Kidney	212
	B. Medullary Anatomy and Concentrating Ability	213
	C. Reabsorption of Solute-free Water	217
	D. Diluting Capacity of the Kidney	221
III.	ADH Secretion	221
	A. Response to Osmotic Stimuli	221
	B. Osmosensitive Receptors	224
	C. Response to Redistribution or Alteration of Blood Volume	225
	D. Volume-sensitive Receptors	228
References		228

I. INTRODUCTION

All the primate species are terrestrial. None have adapted to living in fresh or salt water, neither do any occupy desert areas for any length of time. Yet the osmoregulatory mechanisms that primates possess do permit them to inhabit a wide range of environments within these extremes. These mechanisms also permit man to indulge in the psychological and social pleasures associated with drinking various fluids.

As in most other areas of physiology, our knowledge of the osmoregulatory system is based strongly on research in animals other than primates. To a large extent views on the function of this system in primates are based on logical inferences drawn from research on laboratory animals and on fragments of information that are often indirect and limited by the cost and difficulty of studying subhuman primates and by respect for the integrity of the human organism. Because of these limitations this chapter is largely restricted to the two areas of the subject of osmoregulation in which there is extensive information relating specifically to primates, that is, the role of the kidney in water excretion and the stimuli that alter the concentrations of antidiuretic hormone in blood. The control of Na excretion is not discussed because it is inextricably intermeshed with the complex subjects of extracellular fluid volume regulation and blood pressure regulation.

II. THE ROLE OF THE KIDNEY IN OSMOREGULATION

A. STRUCTURE OF THE KIDNEY

The gross structure of the kidney varies among the primates much as it does among the other mammalian species. Oliver (1968) distinguishes two chief anatomical forms of mammalian kidneys: the simple unipapillary type composed of a single renculus and the multirenculated type. In a subvariety of the former the papillary ridge is extended to form a crest and the pelvis is extended to form tubi maximi. The renculi in the multirenculated kidney may be individually discrete or fused into a single compound mass enclosed by a continuous cortex. As Sperber (1944) has shown, the distribution of these types of kidneys among the mammals is not related to orders or to environmental habitat but seems to be entirely random. The simple unipapillary kidney may be found in both Monotremata and Primata, in both kangaroo rats and horses. The discrete multirenculated kidney occurs in whales, seals and bears among others and the compound multirenculated kidney may be found in beavers, pigs and men. Among primates, the unipapillary kidney evidently prevails with the exception of man and the possible exception of *Ateles* and some of the anthropoid apes.

The cellular anatomy of the primate nephron exhibits only subtle differences from that of the most frequently studied species, the rat. The general shape of the rat and human proximal tubular cells are quite similar. However, the proximal tubular basement membrane in the human is much thicker than in the rat and is often laminated (Tisher *et al.*, 1966). The proximal tubular cell of the rhesus monkey is nearly identical to that of man (Tisher *et al.*, 1969).

The cells of the thin limb of the loop of Henle in the human are quite different from those in the rat, mouse and rabbit. The cell shape is not as complex but the cell height is greater and the cytoplasm contains more abundant rough surfaced endoplasmic reticulum and lipofuscin pigment. The basement membrane is also thicker (Bulger et al., 1967). The transition from the thin limb to the thick limb is gradual in man but abrupt in the rat. In the human, there is an abrupt transition from the thick limb to the columnar cells of the macula densa.

In the rat distal tubule the zonula occludens is longer and the zonula adherens is less well developed than in the proximal tubule. However in man there is little difference in the appearance of these structures between the proximal and distal tubules (Tisher et al., 1968).

Myers et al. (1966) describe four segments of the human collecting ducts: initial, medullary ray, outer medullary and inner medullary. The initial segment is located in the pars convoluta of the cortex and the medullary ray segment in the pars radiata. There are two types of cells in the initial collecting duct: light cells with pale cytoplasm and few organelles interspersed with dark cells having a denser cytoplasm and more numerous organelles. The basilar interdigitations are smaller and less frequent than in the distal tubule. Short lateral infoldings are seen and the mitochondria are less elongate, smaller and not as concentrated in the base of the cell. In the more distal segments the cells progress from cuboidal to low columnar to tall columnar. Dark cells are fairly numerous in the medullary ray segment but become much less so in the more distal segments. Mitochondria and basilar and lateral infoldings become less frequent but small microvilli occur more often. In the rat collecting duct the cells are never higher than low columnar, the basement membrane is not as wide and cilia and desmosomes occur less frequently. However, there are more basilar cell interdigitations and lateral membrane interlocking (Myers et al., 1966).

B. MEDULLARY ANATOMY AND CONCENTRATING ABILITY

Of particular interest is the variety of medullary architecture found among the primates. In some species, not all the nephrons enter the medulla. Oliver (1968) states that these cortical nephrons occur only in compound multirenculated kidneys such as that of man in which 80 to 90% of the nephrons are of this type. However, Sperber (1944) observed cortical nephrons in the unipapillary kidneys of the lemur, the macaque monkey and the baboon. The inner zone of the medulla is very poorly developed in at least four species of the macaque monkey: the rhesus monkey, *Macaca mulatta*; the crab-eating monkey, *M. irus*; the stump-tailed monkey, *M. speciosa* and the pigtailed monkey, *M. nemistrina* (Tisher et al., 1972). Sperber (1944) found the same to be true for the *M. sylvanus* and also the baboon and the lemur. However, man possesses a well-developed inner medulla as does the lemur-like

Chriromys and *Tarsius* (Sperber, 1944); the subprimate senegal bushbaby, *Galago senegalensis senegalensis*; the grivet monkey, *Cercopithecus aethiops aethiops* (Munkacsi and Palkovits, 1966); the squirrel monkey, *Saimiri sciureus* (Selkurt and Wathen, 1967); the owl monkey, *Aotes trivirgatus* (E. E. Selkurt, personal communication) and the chimpanzee, *Pan troglodytes* (Tisher, 1970).

The relationship between the medullary architecture and urine-concentrating ability has been considered by a number of researchers. Sperber (1944) has shown that animals living in an arid habitat in general possess kidneys with a relatively thicker medulla than animals living in an aquatic or wet habitat. Sperber was writing at a time when very little was known about the urine-concentrating mechanism. However, on the basis of his data on the comparative anatomy of the kidney, he rejected an earlier theory that the thin segment of the loop of Henle is the site where the urine is concentrated and suggested instead that the thick segment may be involved. It is known now of course that the thick segment contains the active mechanism that begins the process of urine concentration although it is the urine in the collecting tubule that is concentrated by the process.

Schmidt-Nielsen and O'Dell (1961) later showed a correlation in a variety of animals including man between relative medullary thickness and maximum urine osmotic concentration. They also demonstrated lack of correlation between the fractional number of nephrons with long loops and concentrating ability.

Information on the medullary anatomy and concentrating ability is available in four genera of primates, the macaque, squirrel and owl monkeys, and man. Tisher and co-workers (1972) have studied the medullary anatomy and the urine concentration process of three species of *Macaca*, the rhesus monkey, the crab-eating monkey and the stump-tailed monkey. All three species produce urine averaging 1100–1400 mosmol kg H_2O^{-1} after up to 48 h of water deprivation. As mentioned above, the inner medulla of these species is poorly developed. Bennett *et al.* (1968) found that all the superficial cortical nephrons in the rhesus monkey have loops of Henle that are short and confined to the cortex. The cortex to medulla thickness ratio in the above three species is respectively 0·79, 0·94 and 1·13. The inner medulla comprises 21, 21 and 25% of the total thickness of the medulla (Tisher *et al.*, 1972).

The poorly developed inner medulla attests to the virtual absence of long loops of Henle and in particular to the absence of long thin segments. In the outer medulla, the hairpin turn of the loop is usually formed by epithelium characteristic of the thick ascending limb. The thin limb when present is very short and occurs only in the descending loop. A thin ascending segment is only rarely seen. Tisher and his associates also point out that the outer medulla lacks the organization seen in the outer medulla of other mammals that possess a well-developed inner medulla (Kriz and Lever, 1969). The vascular components are not arranged in bundles but rather are randomly dispersed. Neither is there any obvious pattern in the dispersion of the tubular elements.

The squirrel monkey possesses a concentrating ability similar to that of the macaque (Selkurt and Wathen, 1967; Tanner and Selkurt, 1970). In animals deprived of water for 24 h the urine osmotic pressure averaged 926 mosmol kg H_2O^{-1} with individual values ranging up to 1575 mosmol kg H_2O^{-1}. It seems quite probable that further water deprivation would have led to urine osmolalities in the range observed in the macaque after 48 h of water deprivation. The anatomy of the squirrel monkey kidney is, however, quite different from that of the macaque. The inner medulla is well developed; the cortex to medulla thickness ratio is 0·11 and the inner medulla comprises 75% of the total medullary thickness. In this species all surface nephrons have long loops of Henle extending into the outer medulla (Tanner and Selkurt, 1970).

The owl monkey may have a somewhat smaller concentrating ability. In 17 animals deprived of water for 24 h, the urine osmotic concentration averaged 512 mosmol kg H_2O^{-1}. However, considerably higher values, up to 915 mosmol kg H_2O^{-1}, were obtained from three monkeys of this group. This species also has a well-developed medulla. The cortex to medulla thickness ratio is 0·71 and the inner medulla to total medulla thickness ratio is 0·70 (E. E. Selkurt, personal communication).

In man the maximum osmotic urine to plasma ratio is about 4·0 after 12 to 24 h water deprivation (Wesson, 1969). Lindeman et al. (1960) found the maximum urine osmolality to average 1189 mosmol kg H_2O^{-1} in 14 young men after 24 h of water deprivation. The cortex to medulla thickness ratio is 0·34 and the inner medulla comprises 65% of the total medulla (Sperber, 1944). The loops of Henle in the medulla arise mostly from juxtamedullary nephrons but 80 to 90% of the nephrons are cortical nephrons in which the thin segment is very short and the loop of Henle is comprised mainly of the thick segment which also forms the bend of the loop. These loops pass only a short distance into the outer medulla or may fail to reach the medulla and lie in the medullary rays (Wesson, 1969). Thus the concentrating ability of the macaque, the squirrel monkey and man are quite similar and the owl monkey's ability is somewhat less. There are differences in medullary architecture among the first three with that of the macaque's being most striking.

H

The medullary anatomy of the macaque is quite similar to that of two other species, the mountain beaver, *Aplodontia rufa* and the rodent, *Ondatra zibethicus* (Tisher et al., 1972). However, the macaque can concentrate urine to a much greater extent than either of these species. Because of this Tisher and his associates take issue with Sperber's method of calculating relative medullary thickness and the correlation between this value and concentrating ability alluded to by Sperber (1944) and specifically stated by Schmidt-Nielsen and O'Dell (1961). Sperber defined relative medullary thickness as 10 × medullary thickness × (cube root of the product of the kidney dimensions)$^{-1}$. Tisher et al. (1972) point out that the values for man, 3, and the macaque, 3·1, do not take into account the difference in development of the inner medulla between the two species, as indicated by the ratio of cortex to medulla thickness.

They also draw attention to another discrepancy in the correlation between relative medullary thickness and concentrating ability. The value for the mountain beaver is 2·9 (Sperber, 1944), close to that of man and the macaque, yet the maximum urine osmotic concentration achieved by the beaver is only 550 mosmol kg H_2O^{-1} (Schmidt-Nielsen and Pfeiffer, 1970).

In discussing the similarities of the medullary anatomy and the dissimilarities in concentrating ability between the macaque and the mountain beaver, Tisher et al. (1972) suggest several possible physiological variables that might be responsible for the difference in concentrating ability. Among them are differences in medullary blood flow rates and the capability of establishing a greater salt concentration gradient across the epithelium of the thick segment of the loop of Henle. Certainly it is quite probable that differences in concentrating ability among species is related not only to anatomical differences but also to differences in physiological mechanisms involved.

It has been shown for instance that urea contributes importantly to maximal concentrating ability in several species including man (Schmidt-Nielsen and Robinson, 1970; Epstein, 1956) but it does not do so in the mountain beaver (Schmidt-Nielsen and Pfeiffer, 1970). Tisher et al. (1972) believe it is unlikely that this accounts for the differ ence in concentrating ability between the macaque monkey and the mountain beaver. They point out that in both animals the urea concentration is much higher in urine than in medullary tissue water.

It may be profitable to investigate more fully the role of urea in the concentrating mechanism in those species without a well-developed inner medulla, since the inner and outer medullary structures may handle urea quite differently. Cortical and outer medullary segments of the collecting tubule of the rabbit, an animal with a well-developed inner medulla, are quite impermeable to urea (Grantham and Burg, 1966; Schafer and Andreoli, 1972); so is the thick ascending limb (Rocha and Kokko, 1974). The inner medullary segment of the rat collecting tubule is quite permeable to urea and this permeability may be increased by ADH (Gardner and Maffly, 1964; Morgan and Berliner, 1968). Rocha and Kokko (1974) however did not obtain this effect of ADH in the rabbit. In addition there is evidence of active urea transport out of the papillary collecting duct of the rat (Lassiter *et al.*, 1966; Clapp, 1966). Kokko and Rector (1972) have utilized this and other evidence to propose that inner medullary cycling of urea provides the concentrating effect in the inner medulla. In this proposal the role of the thin ascending limb of the loop of Henle, an inner medullary structure, is crucial.

C. REABSORPTION OF SOLUTE-FREE WATER

There are other measures of the effectiveness of the countercurrent mechanism besides maximal concentrating ability. One such is the ability to reabsorb solute-free water, $T^c_{H_2O}$, particularly while excreting large solute loads. In this regard, there are interesting comparisons to be made among man, the macaque monkey, the squirrel monkey, the dog and the rat.

Tisher *et al.* (1972) infused hypertonic saline and ADH into three species of the macaque monkey and found negative values for $T^c_{H_2O}$ (positive free-water clearance) at high osmolar clearance rates (Fig. 1). Selkurt and Wathen (1967) administered hypertonic mannitol to squirrel monkeys and found that $T^c_{H_2O}$ rose, reached a plateau and then declined as osmolar clearance continually increased. However, they did not obtain any negative values for $T^c_{H_2O}$ even at urine flow rates of 25 ml min^{-1} 100 ml^{-1} glomerular filtrate. The addition of ADH to the infusion tended to counteract the fall in $T^c_{H_2O}$ at high osmolar clearances but did not prevent it entirely. In the presence of exogenous ADH $T^c_{H_2O}$ reached a plateau value of 4·6 ml min^{-1} 100 ml^{-1} glomerular filtrate in a range of urine flow rates from 7 to 25 ml min^{-1} 100 ml^{-1} glomerular filtrate.

In man $T^c_{H_2O}$ attains a maximum of 5 to 10 ml min^{-1} (Raisz *et al.*, 1959; Stein *et al.*, 1962; Goldberg *et al.*, 1965). At high rates of solute excretion $T^c_{H_2O}$ tends to fall but negative values have not been reported. However, Goldberg *et al.* (1965) observed that $T^c_{H_2O}$ continued to rise with no evidence of an upper limit when diuresis was produced by infusion of hypertonic saline. This contrasts sharply with the above data on the macaque in which diuresis was produced in the same way (Fig. 1).

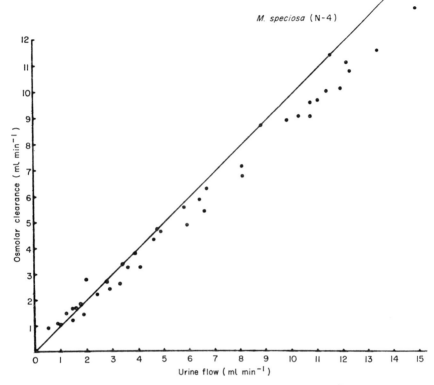

Fig. 1. Relationship between osmolar clearance, C_{osm}, and urine flow, $\dot{V}$, during hypertonic saline infusions. $T^c_{H_2O} = C_{osm} - \dot{V}$, thus $T^c_{H_2O}$ is indicated by the vertical distance of each point from the slope of unity. Points above the line indicate a positive $T^c_{H_2O}$, while points falling below the line represent a negative $T^c_{H_2O}$. Each animal received 50 mU kg^{-1} h^{-1} of vasopressin (from Tisher et al., 1972, published with permission).

In the dog, the maximum $T^c_{H_2O}$ is about 4 to 7 ml min^{-1} 100 ml^{-1} of glomerular filtrate (Page et al., 1952; Anslow and Wesson, 1955; Goldsmith et al., 1961). However, in the dog $T^c_{H_2O}$ can fall to negative values at high rates of solute excretion (Goldsmith et al., 1961; Goodman et al., 1964; Goldberg and Ramirez, 1967). In the rat $T^c_{H_2O}$ continues to rise as solute excretion increases whether mannitol or hypertonic saline is infused although higher values are obtained with the saline. No plateau or maximum limit is seen (Buckalew et al., 1967).

It is difficult to make quantitative comparisons among these specie's because not all the data quoted can be corrected to body or kidney weight or to units of 100 ml^{-1} glomerular filtrate. However, the data in Fig. 1, when compared to the data on the dog, suggest that the macaque's ability to reabsorb solute-free water is much more easily overcome when solute excretion is raised. Thus in regard to this ability these animals may ranks as follows in ascending order: macaque, dog, squirrel monkey, man and rat.

The inability to reabsorb solute-free water when solute excretion is high may be due to incomplete osmotic equilibration of distal tubular fluid, low volume reabsorption in that segment and subsequent rapid passage of a high volume of hypotonic fluid through the collecting duct (Earley et al., 1961; Clapp and Robinson, 1966; Goldberg and Rameriz, 1967). Clapp and Robinson (1966) observed that the distal tubule: plasma osmolality ratio averaged 0·41 in the dog and Bennett et al. (1968) found the same ratio to equal 0·5 in the macaque (Fig. 2). In both the dog and the macaque, approximately 20% of the filtrate remains at the end of the distal tubule (Bennett et al., 1967; Bennett et al., 1968). In contrast distal tubular fluid in the rat becomes isoosmotic to plasma half way along the length of the tubule and only 8 to 10% of the filtrate remains at the end (Gottschalk and Mylle, 1959; Lassiter et al., 1961).

Data obtained in the squirrel monkey is not entirely in agreement with this theory. Osmotic equilibration is incomplete in the distal tubule of this species also (Fig. 3) but it is able to reabsorb solute-free water at high solute excretion rates (Tanner and Selkurt, 1970). However, the osmotic concentration of distal tubular fluid tends to rise along its length (Fig. 3) and only about 11% of the filtrate is left to enter the collecting tubule. Thus the squirrel monkey may be intermediate in this regard among the macaque, the dog and the rat. There is not sufficient data to allow a determination of where man stands in this hierarchy.

In summary, among three genera of monkeys with similar ability to osmotically concentrate the urine, the owl and squirrel monkey have well-developed inner medullas, while the macaque does not. In both the macaque and squirrel monkey incomplete osmotic equilibration of distal tubular fluid occurs in the presence of ADH. However the squirrel monkey reabsorbs a greater fraction of the filtrate in the distal tubule than the macaque and can reabsorb solute-free water at high rates of solute excretion whereas the macaque cannot. Man most closely resembles the squirrel monkey in that he has a well-developed inner medulla and can reabsorb solute-free water when the rate of solute excretion is high. On the basis of these comparisons it is tempting to suggest first that maximal osmotic concentrating ability is not as good a measurement of the effectiveness of the countercurrent mechanism as the ability to continue to reabsorb solute-free water while excreting a large solute load. Secondly, the limited ability of the macaque in this regard may be related to the absence of a well-developed inner medulla.

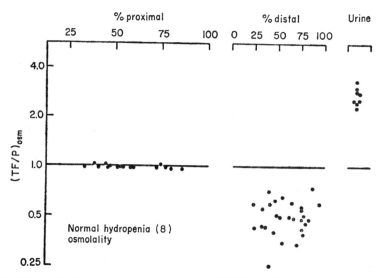

Fig. 2. Tubular fluid to plasma osmolality ratios, $(TF/P)_{osm}$, during hydropenia in eight rhesus monkeys. % proximal and % distal refer to fractional length along each nephron segment (from Bennet et al., 1968, published with permission).

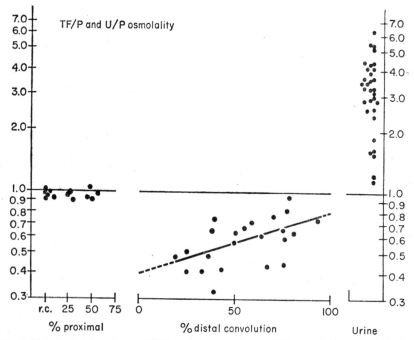

Fig. 3. Tubular fluid to plasma, TF/P, and urine to plasma, U/P, osmolality ratios in hydropenic squirrel monkeys. The regression line indicates the correlation between log $(TF/P)_{osm}$ to % length along the distal convolution. The regression coefficient equals 0·56 ($p < 0.01$) (adapted from Tanner and Selkurt, 1970, published with permission).

D. DILUTING CAPACITY OF THE KIDNEY

There is little information available on the diluting capacity of the kidney in primates other than man. The human kidney is capable of diluting the urine to 40 to 80 mosmol kg H_2O^{-1} (Kleeman et al., 1956; Schoen, 1957). Maximum solute-free water clearance is 10 to 15 ml min^{-1} and this, like solute-free water reabsorption, tends to increase as solute excretion increases, reaching values as high as 28 ml min^{-1} during saline loading (de Wardener and del Greco, 1955; Kleeman et al., 1956).

III. ADH SECRETION

The control of antidiuretic hormone (ADH) secretion in man has been extensively studied and recently some work has been done with other primate species. Investigation in man as in other animals was impeded to some extent by the difficult bioassay methods used to measure ADH levels in plasma (Share, 1967). In many studies inferences as to the direction of change in plasma ADH concentration have been drawn from measurements of urine flow, osmolality or $T^c_{H_2O}$. However, these measurements can be affected by factors other than ADH. Recently very sensitive radioimmunoassay (RIA) methods have been developed for measuring ADH (Robertson et al., 1970; Beardwell, 1971). These have provided a wealth of information already and promise much in the future. Robertson et al. (1977) have extensively reviewed this information.

A. RESPONSE TO OSMOTIC STIMULI

The threshold point for the mechanism regulating plasma osmolality, that is the plasma osmotic concentration below which little or no secretion of ADH occurs, has been determined by RIA methods to average 280 mosmol kg H_2O^{-1} in recumbent men (Robertson et al., 1970, 1973; Robertson, 1974; Robertson and Athas, 1976). At and below this point there is a residual ADH concentration in plasma of 0·5 to 1·0 pg ml^{-1} (1 pg equals 0·4 micro-units of ADH).

The regulating mechanism is so sensitive that a change in plasma osmolality of 1% causes sufficient secretion of ADH to raise the plasma concentration 1 pg ml^{-1} (Fig. 4). This amount is readily detectable and causes a large change in the urine osmotic concentration. Maximum urine concentration can be achieved in man with a plasma

ADH level of 5 pg ml^{-1} (Fig. 5). This is the average plasma concentration that prevails at a plasma osmolality of 295 mosmol kg H$_2$O^{-1} (Fig. 4). Thus the full range of the renal response to ADH from a urine osmolality of 50 to 1200 mosmol kg H$_2$O^{-1} can be obtained by an alteration of plasma osmolality of 15 mosmol kg H$_2$O^{-1} from 280 to 295

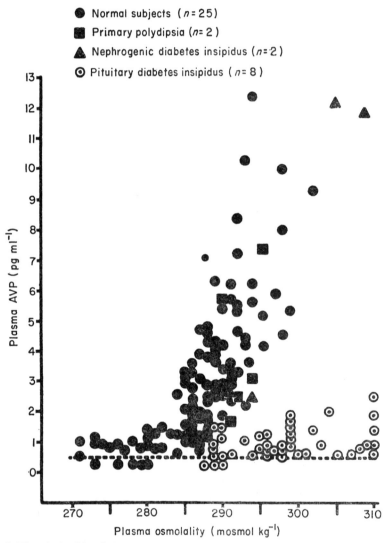

Fig. 4. The relationship of plasma concentration of arginine vasopressin (AVP or ADH) to plasma osmolality in normal subjects and in patients with polyuria of diverse aetiologies. The subjects were recumbent and in various states of water balance. The horizontal broken line indicates the sensitivity limit of plasma vasopressin assay (from Robertson et al., 1973, published with permission).

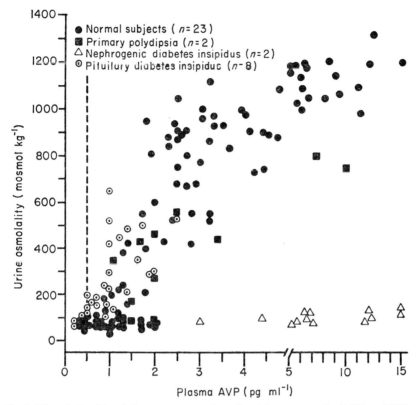

Fig. 5. The relationship of plasma concentration of arginine vasopressin (AVP or ADH) to urine osmolality in normal subjects and patients with polyuria of diverse aetiology. The vertical broken line indicates the sensitivity limit of plasma vasopressin assay (from Robertson et al., 1973, published with permission).

mosmol kg H_2O^{-1} (Robertson et al., 1973; Robertson, 1974). Consider a man weighing 70 kg in which 60% of body weight is water and with a plasma osmotic concentration in the middle of this range. He need lose only 1·1 kg of water or take in 0·93 kg to cause his kidneys to maximally dilute or maximally concentrate the urine. These represent less than a 1·6% change in total body water content.

Various measurements of ADH levels have been made in men in the basal state (ad lib fluid intake) after variable periods of fluid deprivation and in the hydremic state. In one RIA study of recumbent men in the basal state, the plasma osmolality averaged 287 mosmol kg H_2O^{-1} and the average plasma concentration of ADH was 2·7 pg ml^{-1} (1·1 μU ml^{-1}) (Robertson et al., 1973). In three bioassay studies the basal levels of ADH were 1·9, 1·8 and 0·7 μU ml^{-1} (Yoshida et al., 1963;

Czaczkes et al., 1964; Segar and Moore, 1968). In two RIA studies of men deprived of water 8 to 12 h, the plasma osmolalities averaged 288 and 292 mosmol kg H_2O^{-1} and the plasma ADH concentrations were 5·4 and 5·5 pg ml^{-1} (2·2 μU ml^{-1}) (Beardwell, 1971; Robertson et al., 1973). The results of bioassay studies of hydropenic men vary. Share et al. (1972) found a value of 1·8 μU ml^{-1}. Others report values ranging from 4·6 to 8·0 μU ml^{-1}. Plasma osmolalities were not reported in the bioassay studies (Yoshida et al., 1963; Czaczkes et al., 1964; Ahmed et al., 1967). Robertson et al. (1973) reported an average osmolality of 282 mosmol kg H_2O^{-1} for hydremic men and the ADH level was 1·1 pg ml^{-1} (0·6 μU ml^{-1}). The bioassay method is unable to detect the presence of ADH in plasma in this condition, perhaps because of the limitation of its sensitivity (Yoshida et al., 1963; Ahmed et al., 1967).

In one study of sitting, restrained *Macaca mulatta* regulation of ADH secretion occurred over a somewhat higher range of plasma osmolality than in man (Hayward et al., 1976). The threshold osmolality was found to average 292 mosmol kg H_2O^{-1}. At this point and below, the residual plasma ADH concentration was 1·2 to 2·5 pg ml^{-1}. In the basal state the osmolality was 298 mosmol kg H_2O^{-1} and the plasma ADH concentration was 4·3 pg ml^{-1}. Fluid deprivation for 24 h raised these values to 307 mosmol kg H_2O^{-1} and 5·8 pg ml^{-1}.

B. OSMOSENSITIVE RECEPTORS

Hayward and Vincent (1970) have studied the activity of single neurons in the supraoptic nucleus and adjacent regions of the hypothalamus of trained unanaesthetized rhesus monkeys while injecting solutions of varying tonicity into the common carotid artery. They found two major groups of osmo-sensitive cells. The "specific" cells responded to osmotic stimuli but did not respond to non-noxious, arousing, sensory stimuli of sound, light or touch. "Non-specific" cells responded equally to both types of stimuli and were located diffusely throughout the anterolateral hypothalamus. Neither types of cells responded to injections of isotonic saline or distilled water.

There were two subtypes of "specific" cells. Cells in the perinuclear zone of the supraoptic nucleus responded to osmotic stimuli with monophasic accleration or inhibition. The authors propose that these are osmoreceptors on the basis of location, pattern and specificity of response. These data support the earlier suggestions on location of these receptors in the rabbit by Cross and Green (1959) and in cats by Joynt (1964). Cells of the supraoptic nucleus itself showed a biphasic response to osmotic stimuli of acceleration followed by inhibition. The authors consider these to be the secretory neurons of the nucleus.

C. RESPONSE TO REDISTRIBUTION OR ALTERATION OF BLOOD VOLUME

There is evidence, some of it conflicting, that various manoeuvres which tend to either reduce blood volume or shift it from one part of the body to another may alter the rate of ADH secretion. This evidence has been the subject of several recent reviews (Gauer et al., 1970; Share and Claybaugh, 1972; Goetz et al., 1975; Robertson et al., 1977). Much of this evidence is indirect; conclusions in regard to changes in ADH secretion have often been based on alterations in the rate of urine flow and in its composition. This is particularly true of the work with man. Attention will be directed here to experiments in which plasma ADH concentrations have been measured.

Segar and Moore (1968) followed the effect of posture on ADH secretion in eight male subjects. After 60 min in a reclining position, blood (not plasma) concentrations of ADH averaged $0.4\ \mu U\ ml^{-1}$. After 60 min of sitting quitely this rose to $1.4\ \mu U\ ml^{-1}$. The subjects then stood resting against a high stool. The subjects' legs were held motionless and bore little weight. In this position extensive pooling of blood in the legs often results. After 20 min in this position, the blood ADH concentration rose to $3.1\ \mu U\ ml^{-1}$. Share and associates (1972) measured plasma concentrations of ADH in 20 volunteers while they were in bed after 8 h of water deprivation and again after 4 h of ambulation and found no significant differences. In this study, there was probably much less pooling of blood in the legs than in the former study. Robertson and Athar (1976) measured ADH levels in the upright and the recumbent position in 18 subjects, first in the basal state with an average plasma osmolality of 287 mosmol kg H_2O^{-1} and then after 12 to 16 h of fluid deprivation when plasma osmolality averaged 290 mosmol kg H_2O^{-1}. In the basal state the change in posture did not significantly affect the ADH level. In the hydropenic state moving from the upright to the recumbent posture did significantly reduce the ADH concentration.

Negative and positive pressure breathing are known to cause a shift of blood into or out of the thorax. In a variety of studies in dogs and men, continuous positive pressure breathing has been shown to cause an increase in urine flow. Opposite results were obtained with negative pressure breathing (Goetz et al., 1975).

Intermittent positive pressure breathing (IPPB) is the usual method of artificial respiration used clinically with and without the application of a positive end-expiratory pressure (PEEP). Measurements of the effect of these techniques on plasma ADH levels have been made in anaesthetized and conscious patients. Verma et al. (1968) unexpectedly found that plasma antidiuretic activity fell in 13 of 15 anaesthetized subjects when they were subjected to IPPB. Khambatta and Baratz (1972) performed a more detailed study of six conscious patients and found that the plasma ADH concentration rose from 4 to 18 μU ml^{-1}. Plasma osmolality did not change. The authors suggest that anxiety and stress may be the cause of the change in ADH secretion. This may explain the discrepancy between their results and those obtained in anaesthetized patients by Verma et al. (v.s.). However, the authors do not rule out the possibility that a fall in thoracic blood volume may be the cause. Kumar et al. (1974) measured ADH levels in eight patients during IPPB and then during IPPB with PEEP. The plasma ADH concentration was quite high during IPPB, 8·1 μU ml^{-1}, and then rose to 18·8 when IPPB was combined with PEEP.

Patients in various types of stress, in addition to that imposed by artificial respiration, and in various states of hydration do not make ideal subjects for studies of physiological mechanisms. It is obvious that an earlier study on trained normal volunteers could be repeated to some advantage, this time with measurements of ADH concentrations (Murdaugh et al., 1959). In that study continuous positive pressure breathing reduced urine flow, free water clearance, GFR, RPF, solute excretion and Na excretion. Continuous negative pressure breathing increased free water clearance, RPF and solute excretion. The administration of ethanol which inhibits ADH secretion partially reduced the fall in urine flow associated with positive pressure breathing.

Immersion of subjects into water produces an increase in urine flow (Graveline et al., 1962; Behn et al., 1969; Epstein, 1978). Epstein and co-workers (1975) measured the excretion rate of ADH in 10 hydropenic male subjects before, during and after 5 h of immersion to the neck in thermo-indifferent water. During immersion, urine volume, Na and K excretion, creatinine clearance and osmolar clearance increased; $T^c_{H_2O}$ did not change. Urinary excretion of ADH was cut in half and then, after immersion, rose to values higher than those obtained prior to immersion. However, the levels of ADH excretion obtained during immersion were not below the levels measured in normal hydropenic subjects.

Rogge and Moore (1968) have investigated the role of ADH in the antidiuresis associated with application of a partial vacuum to the lower trunk and extremities (lower body negative pressure, LBNP). A pressure of -20 mm Hg had no effect on ADH concentration in blood. A pressure of -30 mm Hg raised the concentration from 1·3 to 3·5 μU ml^{-1} and -40 mm Hg increased the concentration from 1·8 to 4·2 μU ml^{-1}.

Shifts in distribution of blood brought about by gravitational forces have also been shown to alter ADH secretion. Volunteers were subjected to twice the force of gravity in a human centrifuge, first in a posture such that the inertial vector was in parallel to the long axis of the body and acting in a head-to-foot direction. This force causes blood to pool in the lower part of the body. In this situation blood ADH concentrations rose from 1·7 to 4·7 μU ml^{-1}. The application of a positive pressure of 60 mm Hg to the body during centrifugation by an anti-G suit prevented the rise. When the force was applied perpendicular to the long axis of the body in the sternum to spine direction, blood presumably shifted to the thorax and the ADH concentration dropped (Rogge et al., 1967; Moore, 1971).

Plasma ADH levels are often elevated following haemorrhage in laboratory animals (Share, 1967). However, there is a disagreement as to the amount of blood that must be lost to trigger the response (Claybaugh and Share, 1973; Dunn et al., 1973; Goetz et al., 1975). Two studies in man indicate that at least 10% of blood volume must be lost to trigger a rise in ADH secretion (Goetz et al., 1974; Robertson and Athar, 1976). The removal of enough blood from owl monkeys to reduce mean arterial blood pressure to 40 mm Hg did markedly elevate the plasma ADH concentration and retransfusion of the blood reduced the concentration, although not to control levels (Selkurt, 1973). A study of the response of *Macaca mulatta* to haemorrhage indicated that ADH secretion increased only when the rate and degree of haemorrhage was sufficient to cause a fall in arterial pressure. The degree of increase in secretion was not related to the volume of blood withdrawn but rather to the magnitude of the fall in blood pressure and the length of time it was reduced (Arnauld et al., 1977). A later study from the same laboratory (Fumoux et al., 1978) indicated that approximately 10 min periods of hypotension induced by short-term infusions of nitrocyanoferrate (III) produced a rise in ADH secretion. ADH secretion increased exponentially in relation to the degree of fall in arterial pressure.

It is very apparent in all these studies that the loss or redistribution of blood triggers a variety of reactions, both neural and humoral. Alteration in ADH secretion is but one of these. One cannot always be certain that this is a primary response to specific receptors or part of a general response to alterations in the cardiovascular status of the subject. The evidence in regard to the sensitivity of the reflex mechanism or mechanisms governing this response of ADH is conflicting. However the weight of the evidence does seem to indicate that it is much less sensitive than the osmoregulatory mechanism (for example, see Fig. 1 in Chapter 4, Volume 2). Some of the studies reviewed above suggest this. Quiet standing, but not ambulation, increased plasma ADH concentration. LBNP of −20 mm Hg did not raise plasma ADH concentration but −30 mm Hg did. Removal of at least 10% of the blood from humans or monkeys is required to stimulate ADH secretion. Robertson *et al.* (1977) consider that the evidence obtained in studies in both rat and man indicates that haemodynamic changes shift the threshold or "set" of the osmoreceptor control of ADH secretion. Thus a blood volume loss of 10% or more causes a rise in ADH secretion but does not alter the ability of the osmotic control system to respond to changes in blood osmolality.

D. VOLUME-SENSITIVE RECEPTORS

The location of specific receptors that may govern this volume response of ADH has been the subject of much research. The evidence available from primate studies can offer little in this regard. A variety of evidence obtained in other mammals has been reviewed elsewhere in this text (see Chapter 4, Volume 2), by Gauer *et al.* (1970), by Share and Claybaugh (1972) and by Gilmore and Zucker (1975). It is generally believed that the type B receptors located in the left atrium, which respond to changes in atrial distension, give rise to neural impulses which inhibit ADH release. However, this thesis has been sharply questioned by Goetz and co-workers (1975) mainly on the basis that the majority of experimental manoeuvers employed to study the response of these receptors cause haemodynamic changes elsewhere in the body that may trigger other receptors.

Atrial stretch receptors are present in the monkey (Miller and Kasahara, 1964) and in the human (Abraham, 1965). Zucker and Gilmore (1975) studied the responsiveness of type B receptors in the left atrium of rhesus monkeys by recording impulses in the afferent nerves while expanding the animal's blood volume. They found that the receptors in the monkey were much less sensitive to increases in left atrial pressure than the same receptors in the dog (Gilmore and Zucker, 1974). The authors speculate that the difference may be related to the evolution of an upright posture and that a less sensitive atrial receptor mechanism may be more appropriate for the control of blood volume in the primate in which frequent changes in posture are very common (Gilmore and Zucker, 1975).

REFERENCES

Abraham, A. (1965), Microscopic Innervation of the Heart and Blood Vessels in Vertebrates Including Man." Pergamon Press, Oxford.
Ahmed, A. B. J., George, B. C., Gonzales-Anvert, C. and Dengman, J. F. (1967). Increased plasma arginine vasopressin in clinical adrenocortical insufficiency and its inhibition by glucosteroids. *J. clin. Invest.* **46**, 111–123.

6. PRIMATES

Anslow, W., Jr. and Wesson, G., Jr. (1955). Effect of sustained, graded urea diuresis on water and electrolyte excretion. *Am. J. Physiol.* **180**, 605–611.

Arnauld, E., Czernichow, P., Fumoux, F. and Vincent, J. D. (1977). The effects of hypotension and hypovolaemia on the liberation of vasopressin during haemorrhage in the unanaesthetized monkey (*Macaca mulatta*). *Pflügers Arch.* **371**, 193–200.

Beardwell, C. G. (1971). Radioimmunoassay of arginine vasopressin in human plasma. *J. clin. Endocrin. Metab.* **33**, 254–260.

Behn, C., Gauer, O. H., Kirsch, K. and Eckert, P. (1969). Effect of sustained intrathoracic vascular distension on body fluid distribution and renal excretion in man. *Pflugers Arch.* **313**, 123–135.

Bennett, C. M., Clapp, J. R. and Berliner, R. W. (1967). Micropuncture study of the proximal and distal tubule in the dog. *Am. J. Physiol.* **213**, 1254–1262.

Bennett, C. M., Brenner, B. M. and Berliner, R. W. (1968). Micropuncture study of nephron function in the rhesus monkey. *J. clin. Invest.* **47**, 203–216.

Buckalew, V. M., Jr., Ramirez, M. A. and Goldberg, M. (1967). Free water reabsorption during solute diuresis in normal and potassium depleted rats. *Am. J. Physiol.* **212**, 381–386.

Bulger, R. E., Tisher, C. C., Myers, C. H. and Trump, B. F. (1967). Human renal ultrastructure. II. The thin limb of Henle's loop and the interstitium in healthy individuals. *Lab Invest.* **16**, 124–141.

Clapp, J. R. (1966). Renal tubular reabsorption of urea in normal and protein-depleted rats. *Am. J. Physiol.* **210**, 1304–1308.

Clapp, J. R. and Robinson, R. R. (1966). Osmolality of distal tubular fluid in the dog. *J. clin. Invest.* **45**, 1847–1853.

Claybaugh, J. R. and Share, L. (1973). Vasopressin, renin and cardiovascular responses to continuous slow hemorrhage. *Am. J. Physiol.* **224**, 519–523.

Cross, B. A. and Green, J. D. (1959). Activity of single neurons in the hypothalamus: effect of osmotic and other stimuli. *J. Physiol.* (*Lond.*) **148**, 554–569.

Czaczkes, J. W., Kleeman, C. R. and Koenig, M. (1964). Physiologic studies of anti-diuretic hormone by its direct measurement in human plasma. *J. clin. Invest.* **43**, 1625–1640.

DeWardener, H. E. and del Greco, F. (1955). The influence of solute excretion rate on the production of a hypotonic urine in man. *Clin. Sci.* **14**, 715–723.

Dunn, J. L., Brennan, T. J., Nelson, A. E. and Robertson, G. L. (1973). The role of blood osmolality and volume in regulating vasopressin secretion in the rat. *J. clin. Invest.* **52**, 3212–3219.

Earley, L. E., Kahn, M. and Orloff, J. (1961). The effect of infusions of chlorothaizide on urinary dilution and concentration in the dog. *J. clin. Invest.* **40**, 857–866.

Epstein, M. (1978). Renal effects of head-out water immersion in man: Implications for an understanding of volume homeostasis. *Physiol. Rev.* **58**, 529–581.

Epstein, F. H., Kleeman, C. R., Pursel, S. and Hendrikx, A. (1957). The effect of feeding protein and urea on the renal concentrating process. *J. clin. Invest.* **36**, 635–641.

Epstein, M., Pins, D. S. and Miller, M. (1975). Suppression of ADH during water immersion in normal man. *J. Appl. Physiol.* **38**, 1038–1044.

Fumoux, F., Czernichow, P., Arnauld, E., duPont, J. and Vincent, J. D. (1978). Effect of hypotension induced by sodium nitrocyanoferrate (III) on the release of arginine-vasopressin in the unanaesthetized monkey. *J. Endocr.* **78**, 449–450.

Gardner, K. D., Jr. and Maffley, R. H. (1964). An in vitro demonstration of increased collecting tubular permeability to urea in the presence of vasopressin. *J. clin. Invest.* **43**, 1968–1975.

Gauer, O. H., Henry, J. P. and Behn, C. (1970). The regulation of extracellular fluid volume. *Ann. Rev. Physiol.* **32**, 547–595.

Gilmore, J. P., Zucker, I. H. (1974). Discharge of type B atrial receptors during changes in vascular volume and depression of atrial contractility. *J. Physiol. (Lond.)* **239**, 202–233.

Gilmore, J. P. and Zucker, I. H. (1975). The contribution of atrial stretch receptors to salt and water homeostasis in the human. *Basic Research in Cardiology*, **70**, 355–363.

Goetz, K. L., Bond, G. C. and Smith, W. E. (1974). Effect of moderate hemorrhage in humans on plasma ADH and renin. *Proc. Soc. exp. Biol. Med.* **145**, 277–280.

Goetz, K. L., Bond, G. C. and Bloxham, D. D. (1975). Atrial receptors and renal function. *Physiol. Rev.* **55**, 157–205.

Goldberg, M. and Ramirez, M. A. (1967). Effects of saline and mannitol diuresis on the renal concentrating mechanism in dogs: alterations in renal tissue solutes and water. *Clin. Sci.* **32**, 475–493.

Goldberg, M., McCurdy, D. K. and Ramirez, M. A. (1965). Differences between saline and mannitol diuresis in hydropenic man. *J. clin. Invest.* **44**, 182–192.

Goldsmith, C., Beasley, H. K., Whalley, P. J., Rector, F. C. and Seldin, D. W. (1961). The effect of salt deprivation on the urinary concentrating mechanism in the dog. *J. clin. Invest.* **40**, 2043–2052.

Goodman, B., Cohen, J. A., Levitt, M. F. and Kahn, M. (1964). Renal concentration in the normal dog: effect of an acute reduction in salt excretion. *Am. J. Physiol.* **206**, 1123–1130.

Gottschalk, C. W. and Mylle, M. (1959). Micropuncture study of the mammalian urinary concentrating mechanism: evidence for the countercurrent hypothesis. *Am. J. Physiol.* **196**, 927–936.

Grantham, J. J. and Burg, M. B. (1966). Effect of vasopressin and cyclic AMP on permeability of isolated collecting tubules. *Am. J. Physiol.* **211**, 255–259.

Graveline, D. E., Duane, E. and Jackson, M. M. (1962). Diuresis associated with prolonged water immersion. *J. Appl. Physiol.* **17**, 519–524.

Hayward, J. N., Pavasuthipaisit, K., Perez-Lopez, F. R. and Sofroniew, M. V. (1976). Radioimmunoassay of arginine vasopressin in rhesus monkey plasma. *Endocr.* **98**, 975–981.

Hayward, J. N. and Vincent, J. D. (1970). Osmosensitive single neurons in the hypothalamus of unanaesthetized monkeys. *J. Physiol. (Lond.)* **210**, 947–972.

Joynt, R. J. (1964). Functional significance of osmosensitive units in the anterior hypothalamus. *Neurology* **14**, 582–590.

Khambatta, H. J. and Baratz, R. A. (1972). IPPB, plasma ADH, and urine flow in conscious man. *J. Appl. Physiol.* **33**, 362–364.

Kleeman, C. R., Epstein, F. H. and White, C. (1956). The effect of variations in solute excretion and glomerular filtration on water diuresis. *J. clin. Invest.* **35**, 749–756.

Kokko, J. P. and Rector, F. C., Jr. (1972). Countercurrent multiplication system without active transport in inner medulla. *Kidney Intern.* **2**, 214–223.

Kriz, W. and Lever, A. F. (1969). Renal countercurrent mechanism: structure and function. *Am. Heart J.* **78**, 101–118.

Kumar, A., Pontoppidan, H., Bacatz, R. A. and Laver, M. B. (1974). Inappropriate response to increased plasma ADH during mechanical ventilation in acute respiratory failure. *Anesthesiology* **40**, 215–221.

Lassiter, W. E., Gottschalk, C. W. and Mylle, M. (1961). Micropuncture study of net transtubular movement of water and urea in nondiuretic mammalian kidney. *Am. J. Physiol.* **200**, 1139–1146.

Lassiter, W. E., Mylle, M. and Gottschalk, C. W. (1966). Micropuncture study of urea transport in rat renal medulla. *Am. J. Physiol.* **210**, 965–970.

Lindeman, R. D., VanBuren, H. C. and Raisz, L. G. (1960). Osmolar renal concentrating ability in healthy young men and hospitalized patients without renal disease. *New Eng. J. Med.* **262**, 1306–1309.

Miller, M. R. and Kasahara, M. (1964). Studies on the nerve endings in the heart. *Am. J. Anat.* **115**, 217–234.

Moore, W. W. (1971). Antidiuretic hormone levels in normal subjects. *Fed. Proc.* **30**, 1387–1394.

Morgan, T. and Berliner, R. W. (1968). Permeability of the loop of Henle, vasa recta and collecting duct to water, urea and sodium. *Am. J. Physiol.* **215**, 108–115.

Munkacsi, I. and Palkovits, M. (1966). Study on the renal pyramid, loops of Henle and percentage distribution of their thin segments in mammals living in desert, semi-desert and water-rich environment. *Acta biol. acad. sci. hungar.* **17**, 89–104.

Murdaugh, H. V., Jr., Sieker, H. O. and Manfredi, F. (1959). Effect of altered intrathoracic pressure on renal hemodynamics, electrolyte excretion and water clearance. *J. clin. Invest.* **38**, 834–842.

Myers, C. H., Bulger, R. E., Tisher, C. C. and Trump, B. F. (1966). Human renal ultrastructure. IV. Collecting duct of healthy individuals. *Lab. Invest.* **15**, 1921–1950.

Oliver, J. (1968). "Nephrons and Kidneys." Harper and Rowe, New York.

Page, L. B. and Reen, G. H. (1952). Urinary concentrating mechanism in the dog. *Am. J. Physiol.* **171**, 572–577.

Raisz, L. G., Au, W. Y. W. and Scheer, R. L. (1959). Studies on the renal concentrating mechanism. IV. Osmotic diuresis. *J. Clin. Invest.* **38**, 1725–1732.

Robertson, G. L. (1974). Vasopressin in osmotic regulation in man. *Ann. Rev. Med.* **25**, 315–322.

Robertson, G. L. and Athar, S. (1976). The interaction of blood osmolality and blood volume in regulating plasma vasopressin in man. *J. clin. Endocr. Metab.* **42**, 613–620.

Robertson, G. L., Klein, L. A., Roth, J. and Gorden, P. (1970). Immunoassay of plasma vasopressin in man. *Proc. Nat. Acad. Sci. USA* **66**, 1298–1305.

Robertson, G. L., Klein, L. A., Roth, J. and Gorden, P. (1973). Development and clinical application of a new method for the radio-immunoassay of arginine vasopressin in human plasma. *J. clin. Invest.* **52**, 2340–2352.

Robertson, G. L., Athar, S. and Shelton, R. L. (1977). Osmotic control of vasopressin function. In *Disturbances in Body Fluid Osmolality* (Andreoli, T. E., Grantham J. J., Rector, F. C., Jr., eds). American Physiological Society, Bethesda.

Rocha, S. and Kokko, J. P. (1974). Permeability of medullary nephron segments to urea and water: effect of vasopressin. *Kidney Int.* **6**, 379–387.

Rogge, J. D. and Moore, W. W. (1968). Influence of lower body negative pressure on peripheral venous ADH levels in man. *J. Appl. Physiol.* **25**, 134–138.

Rogge, J. D., Moore, W. W., Segar, W. E. and Fasola, A. F. (1967). Effect of +Gz and +Gx acceleration on peripheral venous ADH levels in humans. *J. Appl. Physiol.* **23**, 870–874.

Schafer, J. A. and Andreoli, T. E. (1972). The effect of antidiuretic hormone on solute flows in mammalian collecting tubules. *J. clin. Invest.* **51**, 1279–1286.

Schmidt-Nielsen, B. and O'Dell, R. (1961). Structure and concentrating mechanism in the mammalian kidney. *Am. J. Physiol.* **200**, 1119–1124.

Schmidt-Nielsen, B. and Pfeiffer, E. W. (1970). Urea and urinary concentrating ability in the mountain beaver. *Am. J. Physiol.* **218**, 1370–1375.

Schmidt-Nielsen, B. and Robinson, R. R. (1970). Contribution of urea to urinary concentrating ability in the dog. *Am. J. Physiol.* **218**, 1363–1369.
Schoen, E. J. (1957). Minimum urine total solute concentration in response to water loading in normal man. *J. Appl. Physiol.* **10**, 267–270.
Segar, W. E. and Moore, W. W. (1968). The regulation of antidiuretic hormone release in man. I. Effects of change in position and ambient temperature on blood ADH levels. *J. clin. Invest.* **47**, 2143–2151.
Selkurt, E. E. and Wathen, R. L. (1967). Renal concentrating mechanism of the squirrel monkey. *Am. J. Physiol.* **213**, 191–197.
Selkurt, E. E. (1973). Role of ADH in the loss of renal concentrating ability in primate hemorrhagic shock. *Proc. Soc. exp. Biol. Med.* **142**, 1310–1315.
Share, L. (1967). Vasopressin, its bioassay and the physiological control of its release. *Am. J. Med.* **42**, 701–712.
Share, L. and Claybaugh, J. R. (1972). Regulation of body fluids. *Ann. Rev. Physiol.* **34**, 235–260.
Share, L., Claybaugh, J. R., Hatch, F. E., Jr., Johnson, J. G., Lee, S., Muirhead, E. E. and Shaw, P. (1972). Effects of change in posture and of sodium depletion on plasma levels of vasopressin and renin in normal human subjects. *J. Clin. Endocrin. Metab.* **35**, 171–174.
Shimamoto, K., Murase, T. and Yamaji, T. (1976). A heterologous radioimmunoassay for arginine vasopressin. *J. Lab. clin. Med.* **87**, 338–344.
Sperber, I. (1944). Studies on the mammalian kidney. *Zool. Bid. Fran. Uppsala* **22**, 249–431.
Stein, R. M., Levitt, B. H., Goldstein, M. H., Levitt, M. F., Porrush, J. G. and Eisner, G. M. (1962). Effect of salt restriction on the renal concentrating mechanism in normal hydropenic man. *J. clin. Invest.* **41**, 2101–2111.
Tanner, G. A. and Selkurt, E. E. (1970). Kidney function in the squirrel monkey before and after hemorrhagic hypotension. *Am. J. Physiol.* **219**, 597–603.
Tisher, C. C., Bulger, R. E. and Trump, B. F. (1966). Human renal ultrastructure. I. Proximal tubule of healthy individuals. *Lab. Invest.* **15**, 1357–1396.
Tisher, C. C., Bulger, R. E. and Trump, B. F. (1968). Human renal ultrastructure. III. The distal tubule in healthy individuals. *Lab. Invest.* **18**, 655–668.
Tisher, C. C., Rosen, S. and Osborne, G. B. (1969). Ultrastructure of the proximal tubule of the rhesus monkey kidney. *Am. J. Pathol.* **56**, 469–517.
Tisher, C. C. (1970). Morphology of the kidney: a comparison with man. In "Physiology, Behavior, Serology and Diseases of Chimpanzees" (Bourne, G. H., ed.) Vol. 2. S. Karger, New York.
Tisher, C. C., Schrier, R. W. and McNeil, J. S. (1972). Nature of the urine concentrating mechanism in the macaque monkey. *Am. J. Physiol.* **223**, 1128–1137.
Verma, Y. S., Gupta, K. K., Mehta, S. and Chaudhury, R. R. (1968). A study of plasma antidiuretic activity before and during intermittent positive pressure respiration in human subjects. *Indian J. Med. Res.* **56**, 73–77.
Wesson, L. G., Jr. (1969). "Physiology of the Human Kidney." Grune and Stratton, New York.
Yoshida, S., Motohaski, K., Ibayashi, H. and Okinaka, S. (1963). Method for assay of antidiuretic hormone in plasma with a note on the antidiuretic titer of human plasma. *J. Lab. clin. Med.* **62**, 279–285.
Zucker, I. H. and Gilmore, J. P. (1975). Responsiveness of type B atrial receptors in the monkey. *Brain Res.* **95**, 159–165.

Subject Index

A

Acomys cahirinus
  antidiuretic hormone in, 128
  evaporative water loss, 108
  pulmocutaneous water loss, 112
  urine osmalality, 124
  water intake, 98
  water turnover rate, 107
Acomys russatus
  antidiuretic hormone in, 128
  evaporative water loss, 108
  pulmocutaneous water loss, 112
  urine osmolality, 124
  water intake, 97
Adrenocorticosteroids
  in marsupials, 73
  in monotremes, 73, 87
Aldosterone secretion
  in marsupials, 74
  stimulation of, 191
9-Alpha-Fluorohydrocortisone, 151
Ammonia excretion in birds, 9
Ammoperdix heyi nasal secretion, 41
Ammospermophilus leucurus
  drinking habits, 98, 99
  pulmocutaneous water loss, 110
  urine osmalality, 122
  water intake, 100
  water turnover rate, 106
Amphispiza belli
  urine in, 13
  water budget, 20
Amphispiza belli nevademsis, 4
Anas platyrhynchos
  electrolyte excretion, 31
  urine in, 13
  urine production in, 11
  water budget in, 19
Angiotensin
  in carnivores, 157
  receptor site, 161
  thirst and, 165, 168

Anser anser
  water loss, 6
Antelope
  panting in, 199
Anthochaera carunculata
  urine in, 13
Antidiuretic hormone
  action in rodents, 120, 127
  cardiovascular osmoreceptors and, 153
  effect on nasal salt gland secretion in birds, 42
  haemorrhage affecting, 227
  in carnivores, 157, 172, 176
  in marine mammals, 176
  in marsupials, 72
  in primates, 221
    blood volume and, 225
    osmosensitive receptors, 224
    response to stimuli, 221
    volume sensitive receptors and, 228
  in rabbit, 217
  low pressure receptors and, 157
  negative and positive pressure breathing and, 225
  noradrenaline affecting, 170
  posture and, 225
  prostaglandin E and, 169
  receptors and, 149
  secretion of, 161
Anus sp.
  electrolyte excretion, 32
Aplodontia rufa
  kidney in, 117
  medullary anatomy, 216
  urea excretion, 126
  urine osmalality, 122, 125
  water intake, 100
Apodemus sylvaticus
  water reabsorption in, 129
ATPase, 193
Atrium
  osmoreceptors in, 153, 228

SUBJECT INDEX

Awassi
  urine osmolality, 200

B

Beaver
  kidney in, 216
  urea in, 216
Behaviour
  osmoregulation and in marsupials, 77
Beta adrenergic agonists
  stimulating thirst centre, 168
Birds
  chloride absorption, 34
  cloacal water resorption, 35
  countercurrent multiplier in kidney, 26
  glomerular filtration in, 27
  ionic balance in, 23
  nasal salt glands in, 36, 44
    adaptive pattern of secretion, 44
    control of secretion, 41
    effect of antidiuretic hormone on, 4
    ion transport in, 43
    nature of secretion, 40
    secretory mechanism, 43
    structure and function, 37
  nasal secretions in, 46
  nitrogen excretion in, 9
  osmotic permabilities, 16
  renal structure and function in, 23
  sodium absorption in, 35
  sodium pump in, 26
  urine excretion in, 26
  urine in
    osmotic concentration, 10, 26
    post-renal modification, 34
  water absorption in, 15
  water budget in, 19
  water loss in, 3
    cutaneous, 7
    evaporative, 3
    respiratory, 7
    through nasal salt gland, 17
  water resorption
    rectum and cloaca, 14
Blood volume
  ADH secretion and, 225
Body composition
  fluids and, 197
Body temperature of marine carnivores, 173

Bos taurus, 198
Bradykinin
  effect on sodium metabolism, 170
Brain
  thirst centre in, 165
Buffalo
  panting in, 199
  sweat glands in, 199
  urine osmolality, 200
  water balance in, 195
  water turnover, 187
Buteo jamaicensis
  nasal salt gland secretion, 48

C

Cacatua roseicapilla
  urine in, 12
Calcium excretion by nasal salt glands in birds, 41
Calomys ducilla
  water turnover rate, 106
Calypte anna
  water loss, 4
Camels
  panting in, 199
  salivation in, 202
  sweat glands in, 199
  urine concentration, 206
  urine osmolality, 200
  water excretion in, 192
  water turnover in, 187
Carnivores, 145–184
  antidiuretic hormone in, 172, 176
  drinking behaviour, 161, 165
  hepatic osmoreceptors in, 150
  kallikrein in, 170
  marine, 173–177
    body temperature, 173
    renal function in, 174
    renin-aldosterone system in, 177
  osmoreceptors in
    central, 148
    high pressure, 156
    low pressure, 153
    general problems, 146
  prostaglandin E in, 169
  renin-angiotensis system in, 172
  thirst in, 165–169
  terrestrial, 147–172
    antiotensin and ADH in, 157
    sodium loss in, 148
    sweat losses in, 147

SUBJECT INDEX

urine in, 148
water loss in, 147
Carotid sinus, 157
Carpodacus mexicanus
urine in, 13, 27
water loss, 4
Castor canadensis
urea excretion, 126
urine osmolality, 123, 125
Castor fiber, 96
Cat
thirst in, 167
Cattle
evaporative water loss, 198
faecal water content, 201
salivation in, 202
sweating in, 203
water balance in, 195
water requirements, 198
water turnover, 187
Cavia apera
urine osmolality, 125
Cavia porcellus
urine osmolality, 125
water turnover rate, 107
Cell osmoregulation, 193
Cercartetus nanus
water loss in, 67
Cerophithecus aethiops aethiops
kidney in, 214
Cetacia
antidiuretic hormone in, 176, 177
Chimpanzee
kidney in, 214
Chinchilla laniger
kidney in, 118
renal function in, 120
urine osmolality, 124, 125
water intake, 102
water turnover rate, 107
Chloride absorption in birds, 34
Chloride excretion in birds, 29, 41, 42
Chordeilis acutipennis
water loss, 5
Citellus tridecemlineatus
lactation in, 132
water intake, 100
Clethrionomys gapperi
faecal water content, 130
pulmocutaneous water loss, 111
water intake, 99, 101

Cloaca
water resorption in birds, 14, 35
Cold
herbivores and, 186
Colinus virginanus
urine in, 12
water budget, 21
water loss, 6
Colius striatus
water loss, 5
Coprophagy in rodents, 128, 129
Coturnix coturnix
cations in urine, 33
water budget, 21
Countercurrent multiplier in bird kidney, 26
Cricetulus griseus
water intake, 101
Cynomus ludovicianus
water intake, 97, 99, 100

D

Dacelo gigas
electrolyte excretion, 31
urine concentration, 28
urine in, 13
Dasyuroides byrnei
water loss in, 67
water turnover, 62
Dasycercus cristicauda
drinking pattern, 79
urine concentration, 70
water loss in, 67
water turnover, 62
Dasyurus viverrinus
adrenocorticosteroids in, 73
Dicrostonyx groenlandicus
water turnover rate, 106
Dicrostonyx troquatus
sodium reabsorption, 129
Diffusion, 189
Dikdik
faecal water content, 201
panting in, 199
urine osmolality, 200
water balance in, 195
Dipodomys
kidney in, 117
Dipodomys agilis
drinking habits, 98
pulmocutaneous water loss, 111
urine osmolality, 122

Dipodomys agilis—contd.
  water intake, 100
Dipodomys agilis agilis
  water intake, 100
Dipodomys deserti
  water intake, 97
  water turnover rate, 106
Dipodomys merriami
  evaporative water loss, 105, 108
  faecal water content, 130
  faecal water loss, 128
  food water, 97
  lactation in, 133
  pulmocutaneous water loss, 110, 115, 116
  urine osmolality, 122, 125
  water turnover rate, 106
Dipodomys microps
  pulmocutaneous water loss, 110
  water turnover rate, 106
Dipodomys ordii
  water intake, 100
Dipodomys spectabilis
  pulmocutaneous water loss, 111, 115
Dogs
  angiotensin and ADH in, 162
  central osmoreceptors in, 152
  thirst centre in, 168
  water reabsorption in, 217, 219
Dolichotis patagomum, 96
Donkey
  electrolyte balance in, 202
  panting in, 199
Drinking behaviour
  in carnivores, 161, 165
  in marsupials, 79
Dromaius novaehollandiae
  electrolyte excretion, 31
  urine concentration in, 28
  urine in, 13

E

Eland
  evaporative water loss, 198
  faecal water content, 201
  panting in, 199
  urine osmolality, 200
  water requirement, 198
Electrolyte balance in herbivores, 202
Environment
  effect on osmoregulation in monotremes, 88

Eremophila alpestris
  water loss, 4
Eremopteryx verticalis
  water budget, 20
  water loss, 4
Erethizon dorsatum
  urine osmolality, 124
Estrilda troglodytes
  water loss, 4
Eutamias merriami
  pulmocutaneous water loss, 110, 114
  water intake, 100
Eutamias minutus
  water intake, 100
Eutamias palmeri
  water turnover rate, 106
Eutamias quadrivittatus
  water intake, 100

F

Faecal water
  in herbivores, 201
  in rodents, 128, 134
Faeces
  water and salt loss in marsupials, 74
Falco sparverius
  electrolyte excretion, 31
Food
  selection of in marsupials, 77
Fur licking in marsupials, 66

G

Galago senegaliensis senegaliensis
  kidney in, 214
Galea musteloides
  urine osmolality, 125
Gallus gallus
  glomerular filtration in, 28
  urine in, 12
  urine production in, 11
  water budget, 21
Gazella granti
  evaporative water loss, 198
Gazella thomsoni
  evaporative water loss, 198
Gazelle
  evaporative water loss, 198
  panting in, 199
  sweat glands in, 199
  urine osmolality, 200
  water balance in, 194, 195

water intake, 186
water requirement, 198
Geococcyx californianus
  electrolyte excretion, 31
  nasal secretion, 41
  urine in, 13
  water loss, 6
Gerbillus campestre
  water turnover rate, 107
Gerbillus dasyurus
  faecal water content, 130
Gerbillus gerbillus
  antidiuretic hormone in, 128
  water intake, 97
  water turnover rate, 107
Glomerular filtration in birds, 27
Glomerulus in marsupials, 67
Goat
  evaporative water loss, 198
  faecal water content, 201
  sweat glands in, 199
  urine osmolality, 200
  water turnover in, 187
Gut role of in marsupials, 74

H

Haemorrhage
  plasma ADH levels and, 227
Hartebeest
  faecal water content, 201
  urine osmolality, 200
  water requirement, 198
Herbivores, 185–209
  aldosterone release in, 191
  body composition and fluid, 197
  cell osmoregulation in, 193
  cutaneous evaporation in, 203
  diet of, 186
  electrolyte balance in, 202
  evaporative water loss in, 198
  faecal water content, 201
  habitat, 185
  kidney in, 189, 191
  milk and osmotic balance in, 204
  mineral deficiency in, 205
  osmoregulation, 188
  panting in, 199, 202
  physiological ecotypes, 206
  potassium in, 192
  potassium loss through skin, 204
  renal water loss, 200
  respiratory water loss, 197

  saliva in, 202
  sodium and potassium intake in, 205
  sodium deficiency in, 192
  sodium excretion in, 202
  sweat glands in, 189, 199
  sweating in, 191, 203
  urine osmolality, 200
  vasopressin action in, 192
  water balance in, 194
  water excretion in, 192
  water reabsorption, 201
  water requirements in heat, 198
  water turnover, 187
Horses
  kidney in, 212
  panting in, 199
  salivation in, 202
  sweat glands in, 199
Hydration in marsupials, 81
Hypothalamus osmoreceptors in, 151
Hydrochoerus hydrochaeries, 96
Hyrax
  panting in, 199
  urine osmolality, 200
Hystrix cristata, 96
  kidney in, 117

I

Impala
  urine osmolality, 200
  water requirement, 198
Ionic balance in birds, 23
Ion transport in nasal salt gland in birds, 43
Isoodon macriurus
  drinking pattern, 79
  fur licking in, 66
  hydration, 82
  skin water loss, 65
  thermal stress affecting, 64
  urine concentration, 70
  water consumption and lactation, 76
  water turnover, 62
Isoodon obesulus
  antidiuretic hormone in, 72
Isoproterenol, thirst and, 168

J

Jaculus jaculus
  antidiuretic hormone in, 128
  faecal water content, 131
  urine osmolality, 124
  water intake, 97

Jaculus orientalis
  faecal water content, 131
  pulmocutaneous water loss, 112

K

Kallidin-10
  effect on sodium excretion, 170
Kallikrein
  in carnivores, 170
Kidney
  diluting capacity in primates, 221
  in herbivores, 189, 191
  in marine carnivores, 174
  in marsupials, 67
  in rodents, 134
  organization in birds, 23
  role in osmoregulation, in primates, 211
  structure of
    in birds, 23
    in primates, 211
    in rodents, 117
  water loss in birds, 9
Kidney function in monotremes, 87
Kobus ellipsiprymmus
  water balance in, 195
Kongoni
  water balance in, 195

L

Lactation
  in rodents, 129
  water turnover and
    in herbivores, 202
Lagopus leucurus
  water logs, 6
Lagostomus maximus
  urine osmolality, 125
Lagurus lagurus
  water intake, 102
Lanius ludovicianus
  water loss, 5
Laphortyx gambelii
  water budget, 21
Larus argentatus
  nasal fluid composition, 40
Leggadina
  kidney in, 117
Leggadina hermannsburgensis
  faecal water content, 131
  pulmocutaneous water loss, 112
  urine osmolality, 124, 125
  water intake, 97, 102

Lemmus lemmus
  faecal water content, 130
  sodium reabsorption, 129
  water intake, 101
Lemur
  medullary anatomy, 213
Lepus oryctolagus
  pulmocutaneous water loss, 115
Liomys irroratus
  drinking habits, 98
  pulmocutaneous water loss, 111
  urine osmolality, 122
  water intake, 100
  water turnover rate, 106
Liomys salvani
  pulmocutaneous water loss, 111
  urine osmolality, 122
  water intake, 100
  water turnover rate, 106
Liver
  osmoreceptors in, 150
Llama
  salivation in, 202
Lochura malabarica
  water loss in, 4, 8
Lophortyx californicus
  urine in, 12
  water loss, 6
Lophortyx gambelli
  glomerular filtration rate in, 28
  urine in, 12
  water loss, 5
Lung
  evaporation from
    in marsupials, 63
  water loss, 190
    in rodents, 108, 109–115

M

Macacca irus
  kidney in, 213
Macaca mulatta
  haemorrhage and ADH in, 227
  kidney in, 213
  plasma osmolality, 224
Macacca sylvanus
  kidney in, 213
Macropus eugenii
  drinking pattern, 79
  faecal water loss, 75
  fur licking in, 66
  hydration, 82

salt in urine, 71
sea water drinking by, 80
thermal stress affecting, 64
urine concentration, 70
urine flow in, 68
urine in, 68
Macropus giganteus
  adrenocorticosteroids in, 73, 74
  sodium in urine, 71
  water consumption and lactation, 76
Macropus robustus
  drinking pattern, 80
  fur licking in, 66
  gut as water store, 74
  migration in, 84
  nutrition, 83
  skin water loss, 65
  thermal stress affecting, 64
  urinary concentration, 82
  urine concentration in, 70
  water availability and reproduction, 83
  water consumption and lactation, 76
Macropus rufa
  sweat glands in, 64
  urine flow in, 68
Macropus rufogriesus
  pulmocutaneous evaporation in, 63, 65
  sweat glands in, 64
Macrotis lagostis
  skin water loss, 65
  thermal stress affecting, 64
  urine concentration, 70
  water loss in, 67
  water turnover, 62
Magnesium excretion in birds, 29
Mammals
  pulmocutaneous water loss, 109
Marsupials
  abundance and survival, 84
  adrenocorticosteroids in, 73
  aldosterone in, 74
  antidiuretic hormone action in, 72
  behaviour and osmoregulation, 77
  body temperature, 58
  composition of milk, 76
  drinking patterns, 79
  drinking sea water, 80
  effects of thermal stress, 64
  effects of water availability on life of, 81
  evaporative water loss in, 78
  faecal water and salt loss, 74
  food selection in, 77
  fur licking in, 66
  habitat, 57–60, 78
  hydration in, 81
  kidney in, 67
  migration in, 84
  newborn, water regulation in, 76
  nutrition in, 83
  osmoregulation, problems of, 61
  renal clearances in, 69
  renal function in, regulation, 72
  reproduction, effect of water availability, 83
  role of gut in, 74
  selection of refuges, 78
  sweat glands in, 64
  urine concentration in, 68, 70, 82
  urine formation in, 67
  water and salt metabolism, effect of reproduction, 75
Megaleia rufa
  adrenocorticosteroids in, 73
  drinking pattern, 79
  fur licking in, 66
  migration in, 84
  response to adrenalectomy, 72
  selection of refuge, 78
  skin water loss, 65
  thermal stress affecting, 64
  urine concentration in, 70
  urine flow in, 68
  water and reproduction, 83
  water intake, 186
  water turnover, 62
  water turnover in, 187
Meleagris gallopave
  electrolyte excretion, 30
  urine in, 12
Meliphaga virescens
  electrolyte excretion, 30
  urine in, 13, 27
Melopsittacus undulatus
  electrolyte excretion, 30
  urine in, 12, 27
  urine production in, 11
  water budget, 20
  water loss, 4
Meriones crassus
  faecal water content, 130
  water intake, 97

Meriones shawi
  kidney in, 118
  urea excretion, 126
  urine osmolality, 123
  water turnover rate, 107
Meriones unguiculatus
  water intake, 102
  water turnover rate, 107
Mesocricetus auratus
  evaporative water loss, 108
  faecal water content, 130
  kidney in, 118
  pulmocutaneous water loss, 111
  water intake, 101
Micronisus gabar
  electrolyte excretion, 31
Microtus
  water turnover rate, 104
Microtus abbreviatus fisheri
  water turnover rate, 107
Microtus agrestis
  water intake, 101
Microtus californicus
  pulmocutaneous water loss, 111
Microtus ochrogaster
  drinking habits, 99
  pulmocutaneous water loss, 111
  water intake, 102
Microtus oecinomus macfarlani
  water turnover rate, 107
Microtus pennsylvanicus
  drinking habits, 99
  pulmocutaneous water loss, 111
  urine osmolality, 123
  water intake, 101
Microtus pennsylvanicus tananaensis
  water turnover rate, 107
Migration in marsupials, 84
Milk
  composition of
    in marsupials, 76
  osmotic balance and
    in herbivores, 204
Mimus polyglottos
  water loss, 5
Mineral deficiency in herbivores, 205
Molothrus ater
  water loss, 5
Monkey
  kidney in, 214
  medullary anatomy, 213, 216
  osmosensitive receptors to ADH, 224
  plasma osmolality, 224
  water reabsorption, 217, 218, 219
Monotremes, 86–90
  adrenocorticosteroid hormones in, 73, 87
  body temperature, 58, 86
  effect of environment on osmoregulation, 88
  evaporative water loss in, 86
  habitat, 58–60
  renal clearances, 69
  renal function in, 87
Motomys alexis
  renal function in, 120
Mus musculus
  antidiuretic hormone in, 128
  coprophagy in, 129
  evaporative water loss, 108
  faecal water content, 131
  kidney in, 117, 118
  lactation in, 132, 133
  pulmocutaneous water loss, 111, 112, 114, 115
  urea excretion, 126
  urine osmolality, 123, 124
  water intake, 99, 102
  water turnover rate, 104, 107
Muzzle glands, 191
Myocaster coypu
  urea excretion, 126
  urine osmolality, 124
Myopis schisticolor
  sodium reabsorption, 129

N

Napaeozapus insignis
  pulmocutaneous water loss, 112
  water intake, 99, 102
Nasal salt glands
  in birds, 36
    adaptive pattern of secretion, 44
    control of secretion, 41
    effect of antidiuretic hormone on, 42
    ion transport, 43
    nature of secretion, 40
  mechanism, 43
    nature of secretion, 44
    sodium chloride secretion by, 36
    structure and function, 37
    water loss through, 17

Neofiber alleni
 urea excretion, 126
 urine osmolality, 123
Neophema bourki
 water loss, 4
Neotoma floridana
 water intake, 99
Neotoma floridana osagensis
 water intake, 101
Neotoma fuscipes
 pulmocutaneous water loss, 111
Neotoma fuscipes macrotus
 water intake, 99, 101
Neotoma lepida
 pulmocutaneous water loss, 111
 water intake, 101
 water turnover rate, 106
Neotoma micropus canescens
 water intake, 101
Nephron
 diffusion and, 190
 in birds, 23, 26
 in primates, 212
Nitrogen excretion in birds, 9
Noradrenaline, effect on urine, 169
Nose
 water loss from in rodents, 115
Notomys
 kidney in, 189
Notomys alexis
 faecal water content, 131
 lactation in, 133
 pulmocutaneous water loss, 112, 116
 urine osmolality, 121, 124, 125
 water intake, 97, 102
 water turnover rate, 104
Notomys cervinus
 faecal water content, 131
 lactation in, 133
 pulmocutaneous water loss, 112
 urine osmolality, 124, 125
 water intake, 99, 102
Nutrition in marsupials, 83
Nymphicus hollandicus
 water loss, 5

O

Oceanodroma leucorrhoa
 nasal secretion, 40
Octodon degus
 urine osmolality, 125

Ocyphaps lophotes
 electrolyte excretion, 30
 urine in, 12
Onchomys torridus
 urine osmolality, 123
Ondatra zibethica
 kidney in, 118, 216
 urea excretion, 126
 water turnover rate, 107
Ondatra zibethica osyoosensis
 urine osmolality, 123
Onychomys leucogaster
 water intake, 101
Onychomys torridus
 pulmocutaneous water loss, 111
 urea excretion, 126
Ornithorhynchus anatinus
 adrenocorticosteroids in, 73
 osmoregulation in, 88
Oryx
 evaporative water loss, 198
 sweat glands in, 199
 urine osmolality, 200
 water balance in, 194, 195
 water intake, 186
 water requirement, 198
Osmoreceptors
 atrial, 153, 228
 cellular, 193, 224
 central, 148
 hepatic, 150
 high pressure, 156
 low-pressure, 153, 228
Osmoregulation
 cellular, 193
 general problems, 146
 meaning of, 146
Osmotic pressure, 188
Otus asio
 water loss, 5
Oxytocin, effect on sodium excretion, 171

P

Panting
 in herbivores, 199, 202
 water loss in, 192
Pan troglodytes
 kidney in, 214
Passer domesticus
 water loss, 4
Passerculus sandwichensis beldingi
 urine in, 13, 27

242 SUBJECT INDEX

Passerculus sandwichensis brooksi
  urine in, 13
Pelecanus erythrorhynchos
  electrolyte excretion, 31
  urine in, 13
Peophila guttata
  urine in, 13, 27
  water loss, 4
Perameles nasuta
  adrenocorticosteroids in, 73
  skin water loss, 65
  thermal stress affecting, 64
  urine concentration, 70
  water turnover, 62
Perognathus baileyi
  pulmocutaneous water loss, 110
Perognathus californicus
  pulmocutaneous water loss, 110
  water intake, 97
Perognathus fallax fallax
  water intake, 99, 100
Perognathus formosus
  water turnover rate, 106
Perognathus intermedius
  pulmocutaneous water loss, 110
  water intake, 97
Perognathus parvus
  pulmocutaneous water loss, 110
Perognathus penicillatus
  pulmocutaneous water loss, 110
Peromyscus californicus insignis
  water intake, 101
Peromyscus crinitus
  pulmocutaneous water loss, 111
  water intake, 97
Peromyscus crinitus stephensi
  urea excretion, 126
  urine osmolality, 123
  water intake, 101
Peromyscus eremicus fraterculus
  water intake, 101
Peromyscus floridanus
  water 101
Peromyscus leucopus
  faecal water content, 130
  pulmocutaneous water loss, 111
  water intake, 99, 101
Peromyscus leucopus novaboracensis
  urine osmolality, 123
Peromyscus maniculatus
  pulmocutaneous water loss, 111, 114
  urine osmolality, 123

  water intake, 98, 101
  water turnover rate, 104
Peromyscus maniculatus bardii
  water turnover rate, 106
Peromyscus maniculatus rufinus
  lactation in, 132
Peromyscus maniculatus sonoriensis
  evaporative water loss, 105
Peromyscus polinotis subgriesus
  water intake, 101
Peromyscus truei
  faecal water content, 130
  urine osmolality, 123
  water intake, 101
  water turnover rate, 105, 106
Petrogale inornata
  hydration, 82
  water consumption and lactation, 76
  water turnover, 62
Phalacrocorax auritus
  electrolyte elimination, 18
Phalaenoptilus nuttalii
  water loss, 5
Phascolarctos cinereus
  adrenocorticosteroids in, 73
Pigs, muzzle glands, 191
Pipilo aberti
  water loss, 5
Pipilo erythrophthalmus
  water loss, 5
Pipilo fuscus
  water loss, 5
Plasma osmolality in primates, 223
Platycercus zonarius
  water loss, 5
Poephila guttata
  electrolyte excretion, 30
  water budget, 20
  water loss in, 8
Porpoise
  antidiuretic hormone in, 176
  respiration in, 173
Posture
  antidiuretic hormone and, 225
Potassium
  dietary, in herbivores, 205
  urinary, in birds, 34
Potassium excess in herbivores, 192
Potassium excretion
  atrial pressure and, 155
  in birds, 29
  in marsupials, 71

SUBJECT INDEX

Potassium loss
  in cutaneous secretions, in herbivores, 204
Potassium secretion
  by nasal salt gland in birds, 41
Primates, 211–232
  antidiuretic hormone secretion in, 221
    blood volume and, 225
    osmosensitive receptors, 224
    response to stimuli, 221
    volume sensitive receptors and, 228
  diluting ability of kidney in, 221
  kidney structure in, 211
  medullary anatomy and concentrating ability, 213
  plasma osmolality, 223
  urea in, 216
  urine osmolality, 215
  water reabsorption, 217
Prostaglandin E in carnivores, 169
Psammomys obesus
  renal function in, 120
  urea excretion, 126
  urine osmolality, 123, 125
  water intake, 102
Pseudomys australis
  lactation in, 133
  pulmocutaneous water loss, 111
  water intake, 102

R

Rabbit
  antidiuretic hormone in, 217
  osmosensitive receptors, 224
Rats
  angiotensin and ADH in, 161
  see also Rattus
  hepatic osmoreceptors in, 151
  nephron in, 212
  water reabsorption in, 219
Rattus
  hepatus osmoreceptors in, 151
Rattus exulans
  reproduction in, 133
Rattus fuscipes
  pulmocutaneous water loss, 111
Rattus lutreolus
  pulmocutaneous water loss, 111
Rattus norvegicus
  evaporative water loss, 108
  faecal water content, 131
  kidney in, 117
  lactation in, 132
  pulmocutaneous water loss, 111, 114
  urea excretion, 126
  water intake, 102
  water reabsorption in, 129
  water turnover rate, 107
Rattus rattus
  pulmocutaneous water loss, 111, 115
  reproduction in, 133
Rattus villosissimus
  pulmocutaneous water loss, 111, 115
Rectum
  water resorption in, in birds, 14
Reithrodontomys megalotis
  water intake, 100
Reithrodontomys megalotis limicola
  water intake, 100
Reithrodontomys megalotis longicauda
  water intake, 101
Reithrodontomys raviventris
  urea excretion, 126
  urine osmolality, 123
Reithrodontomys raviventris haliocetis
  water intake, 101
Renal clearances
  in marsupials, 69
Renal function
  regulation in marsupials, 72
Renal nerve stimulation
  renin release and, 168
Renin
  in marsupials, 72
Renin-aldosterone system
  in marine mammals, 177
Renin-angiotensin system
  in carnivores, 157, 172
Renin release
  renal nerve stimulation and, 168
Reproduction
  effect of water availability on, in marsupials, 83
Respiration
  in marine carnivores, 173
  negative and positive pressure antidiuretic hormone and, 225
Respiratory water loss
  in birds, 7
  in herbivores, 197
Rodents, 95–144
  antidiuretic hormone activity in, 120, 127
  classification, 95

Rodents—*contd.*
  coprophagy in, 128, 129
  cutaneous water loss in, 108
  drinking habits, 98
  evaporative water loss in, 105, 134
  faecal water loss in, 128, 134
  habitat, 96
  kidney in, 134
    sodium chloride transport in, 121
    structure in, 117
  lactation in, 129
  pulmocutaneous water loss, 108
    in various habitats, 113
  sweat glands in, 116
  urea excretion in, 126
  urine concentration in, 120
  urine in, 117–128
  urine osmolality in, 121
  water balance in, 134
  water intake, 97
  water loss from nose, 115
  water turnover rate, 104

S

Saimiri sciureus
  kidney in, 214
Saliva
  in herbivores, 202
Salt metabolism
  effect of reproduction on, in marsupials, 75
Sarcophilus harrisii
  adrenocorticosteroids in, 73
  skin water loss, 65
Scardafella inca
  water loss, 5
Seal
  respiration in, 173
Sea water
  drinking of by marsupials, 80
Setonix brachyurus
  adrenocorticosteroids in, 73
  antidiuretic hormone in, 72
  faecal water loss, 75
  food preferences, 77
  fur licking in, 66
  neurohypophyseal hormones, 72
  pulmocutaneous evaporation from, 63
  salt excretion, 71
  sea water drinking by, 80, 81
  urine concentration, 70
  urine in, 68
  water loss in, 67
Sheep
  evaporative water loss, 198
  faecal water content, 201
  panting in, 192
  salivation in, 202
  sweat glands in, 199
  sweating in, 204
  urine osmolality, 200
  water balance in, 195
  water intake, 186
  water turnover in, 187
Sitta canadensis
  water loss, 4
Skin
  evaporation from, in marsupials, 63
  potassium loss via, in herbivores, 204
  water crossing, 189
  water loss from, 108
    in birds, 7
    in rodents, 109–115
Sminthopsis crassicaudata
  water loss in, 67
  water turnover, 62
Sodium balance
  in terrestrial carnivores, 147
Sodium chloride
  secretion by nasal salt glands, 36, 44
  transport in rodent kidney, 121
Sodium deficiency, 147
  in herbivores, 192, 205
Sodium excess, 147
Sodium excretion
  atrial pressure and, 155
  atrial receptors and, 156
  bradykinin affecting, 170
  in birds, 29
  in herbivores, 202
  in marine mammals, 174
  in marsupials, 71
  in terrestrial carnivores, 148
  oxytocin affecting, 171
  receptors, 155
Sodium pump
  in birds, 26
Sodium resorption
  in birds, 34, 35, 44
  in carnivores, 163
Sodium secretion
  by nasal salt glands in bird, 41, 42
Sodium transport
  angiotensin affecting, 164

SUBJECT INDEX

Speotyto cunicularia
  water loss, 5
Spermophilus armatus
  pulmocutaneous water loss, 110
Spermophilus beecheyi
  evaporative water loss, 108
  pulmocutaneous water loss, 110, 116
Spermophilus columbianus
  urine osmolality, 122
Spermophilus lateralis
  pulmocutaneous water loss, 110
  water turnover rate, 106
Spermophilus richardsoni
  pulmocutaneous water loss, 110
Spermophilus spilsoma
  pulmocutaneous water loss, 110
Spermophilus tereticaudatus
  lactation in, 133
  water turnover rate, 106
Spermophilus townsendi
  pulmocutaneous water loss, 110
Spizocorys starki
  water loss, 4, 8
Stellula caliope
  water loss, 4
Streptopelia senegalensis
  urine in, 12, 27
Struthio camelus
  nasal secretion, 41
  urine in, 13
  water loss, 6
Sweat
  in carnivores, 147
  in herbivores, 191, 203
Sweat glands
  in herbivores, 189, 199
  in marsupials, 64
  in rodents, 116
Syncerus caffer
  evaporative water loss, 198
  water balance in, 195

T

Tachyglossus aculeatus
  adrenocorticosteroids in, 73
  evaporative water loss in, 86
  renal clearance in, 69
  renal function in, 87
  urine concentration, 70
  water balance in, 88, 89
Tamias striatus
  water intake, 99, 100

Tears, 191
Terathopius ecaudatus
  electrolyte excretion, 31
Thirst
  angiotensin and, 165, 168
  in carnivores, 165–169
Thirst centre, 165
  in dogs, 168
  stimulation of, 168
Toxostoma redivium
  water loss, 5
Trichosurus caninus
  antidiuretic hormone in, 72
Trichosurus vulpecula
  adrenocorticosteroids in, 73
  fur licking in, 66
  hydration, 82
  kidney in, 68
  neurohypophyseal hormones, 72
  renal clearances, 69
  response to adrenalectomy, 72
  thermal stress affecting, 64
  urine concentration, 70
  water consumption and lactation, 76
  water turnover, 62
Troglodytes aedon
  water loss, 4

U

Urea
  diffusion, 190
  in primates, 216
Urea excretion
  in birds, 9
  in marsupials, 71
  in rodents, 126
Urine
  concentration
    in birds, 26
    in man, 221
    in marsupials, 68, 70, 82
    in rodents, 120
  excretion in birds, 26
  formation in marsupials, 67
  in marine mammals, 174
  in rodents, 117–128
  in terrestrial carnivores, 148
  osmolality
    in herbivores, 200
    in primates, 215
    in rodents, 121
  osmotic concentration of, in birds, 10

Urine—*contd.*
  post-renal modification of, in birds, 34
  production in birds, 10
Urine flow
  atrial receptors and, 156
  central control of, 151, 153
  negative pressure breathing and, 153
  respiration and, 226

V

Vagus nerves, 157
Vasopressin, 155
  action in herbivores, 192
  decreasing water loss, 115
  in sheep, 193
Vombatus hirsutus
  adrenocorticosteroids, 73, 74
  sodium concentration in urine, 71

W

Water
  availability of
    effect on marsupials, 81
    effect on reproduction, 83
  cutaneous evaporation in herbivores, 203
  faecal content
    in herbivores, 201
    in rodents, 128
Water absorption in birds, 15
Water balance
  in herbivores, 194
  in marine mammals, 173
  in rodents, 134
  pathways and control in birds, 2
Waterbuck
  water balance in, 195
Water budgets in birds, 19
Water consumption in lactating marsupials, 76
Water excretion in herbivores, 192
Water intake
  of herbivores, 186
  of rodents, 97
Water loss
  cutaneous, 189
    in birds, 7
    in rodents, 108
  cutaneous and respiratory compared in birds, 7
  evaporative
    in birds, 3, 7
    in herbivores, 198
    in marsupials, 78
    in monotremes, 86
    in rodents, 105, 134
    rate in birds, 3
  faecal
    in rodents, 134
    in herbivores, 200
    in marsupials, 65
    in terrestrial carnivores, 147
  nasal
    in rodents, 115
  pulmocutaneous, 190
    in birds, 7
    in herbivores, 197
    in rodents, 109–115
  renal, in birds, 9
Water metabolism
  effect of reproduction on, in marsupials, 75
Water resorption
  in cloaca and rectum in birds, 14, 35
  in herbivores, 201
  in primates, 217
  in rodents, 129
Water requirement
  environment and, in rodents, 133
  in herbivores, 198
Water turnover
  in birds, 3
  in herbivores, 187
  marsupials, 61
  in rodents, 104
  lactation and, in herbivores, 202
Whale
  respiration in, 173
Wildebeest
  sweat glands in, 199
  urine osmolality, 200
  water balance in, 195
  water loss, 198
  water requirement, 198

Z

Zebu
  urine osmolality, 200
Zenaida macroura
  urine in, 12, 27
  water loss, 5
Zonotrichia leucophyrys
  water loss, 4

ST. MARY'S COLLEGE OF MARYLAND
ST. MARY'S CITY, MARYLAND 20686